Groundwater Management and Resources

Edited by Bahareh Kalantar

Published in London, United Kingdom

IntechOpen

Supporting open minds since 2005

Groundwater Management and Resources
http://dx.doi.org/10.5772/intechopen.87804
Edited by Bahareh Kalantar

Contributors
Edwin Pino, Andre Steenken, Yao Shan, Sanjeev Pandey, Randall Cox, Steven Flook, Naseem Akhtar, Mohd Talha Anees, Muhammad Izzuddin Syakir Ishak, Mohammad Shaiful Yusuff, Abdul Qadir, Pawan Jeet, Jatoth Veeranna, Venaktesan Madha Suresh, Ibrahim Lawal Lawal Kane

First published in London, United Kingdom, 2021 by IntechOpen
IntechOpen is the global imprint of INTECHOPEN LIMITED, registered in England and Wales, registration number: 11086078, 5 Princes Gate Court, London, SW7 2QJ, United Kingdom
Printed in Croatia

British Library Cataloguing-in-Publication Data
A catalogue record for this book is available from the British Library

Additional hard and PDF copies can be obtained from orders@intechopen.com

Groundwater Management and Resources
Edited by Bahareh Kalantar
p. cm.
Print ISBN 978-1-83881-064-1
Online ISBN 978-1-83881-065-8
eBook (PDF) ISBN 978-1-83881-066-5

For EU product safety concerns:
IN TECH d.o.o., Prolaz Marije Krucifikse Kozulić 3, 51000 Rijeka, Croatia, info@intechopen.com or visit our website at intechopen.com.

Meet the editor

Dr. Bahareh Kalantar is a research scientist in machine learning, remote sensing, image processing, and natural hazard modelling at the RIKEN Center for the Advanced Intelligence Project, Tokyo, Japan. She has a Ph.D. in Remote Sensing from Universiti Putra Malaysia and completed her postdoctoral studies at RIKEN. Dr. Kalantar was a visiting scholar at the University of Technology Sydney, Australia, in April 2019. She was awarded the Australia Endeavour Fellowship in 2018. In 2019, she received the best paper award at the ISPRS Geospatial week in Enschede, the Netherlands. More recently, she received the 11 Ohbu Award (Researcher Incentive) from RIKEN. She has participated in many international conferences and has published several ISI journal papers.

Contents

Preface

Well over 40% of the world's population is affected by water scarcity. The World Bank works with the majority of nations to ensure that water use is sustainable, that climate resilience is built, and that integrated management is strengthened. Some countries are currently putting an unprecedented amount of strain on their water supplies. Water exists in various forms, such as lakes and rivers, glaciers and ice sheets, oceans and seas, subversive aquifers, and fog in air and clouds. The water cycle refers to how water enters and leaves the atmosphere; it evaporates from the Earth's surface and falls back as precipitation. Like waterfalls, it accumulates in bodies of water and replenishes the groundwater level.

The world's population is rapidly growing with forecasts estimating that, if current systems continue, the Earth will confront a 40% gap between estimated demand and accessible water supply by 2030. Recurring water scarcity, hydrological unpredictability, and extreme weather events (floods and droughts) are significant global prosperity and stability challenges. The importance of water scarcity and drought in exacerbating fragility and conflict is becoming better recognized.

By 2050, providing for nine billion individuals will demand a 60% increase in food production (which presently uses 70% of the resource) and a 15% increase in water extraction. Separately from this increasing demand, water is at present in short supply in many world regions. According to statistics, 40% of the world's population lives in water-scarce regions, and this dilemma affects about 14% of global GDP. By 2025, around 1.8 billion individuals will be living in water-scarce areas or countries. For several nations today, water security is a critical issue.

Climate change will aggravate conditions through changing hydrological cycles, resulting in inconsistent global water supplies, while floods and droughts are expected to be more frequent, erratic, and intense. These events will put roughly one billion residents in monsoonal basins and 500 million in deltas at serious risk. Historically, an estimated $120 billion per year is accrued just from property damage. Droughts pose potential subsistence limitations to poor countryside demographics, which are at the mercy of unpredictable rainfall.

The fragmentation of water further hampers water security. A total of 276 transboundary basins are divided by 148 nations, accounting for 60% of the world's freshwater flow. From this, 300 aquifer systems are transboundary, meaning around two billion people depend on global groundwater. Fragmentation concerns are commonly mirrored at the national level, requiring teamwork to ensure adequate water resource management and development schemes for all riparian areas. Nations will need to enhance the way they manage their water resources and related components to handle these complex and interrelated water concerns.

Investors will need to invest in institutional consolidation, information management, and (natural and non-natural) infrastructure development to increase water security in the face of rising demand, water scarcity, growing unpredictability, more significant extremes, and fragmentation concerns. Institutional mechanisms such as legal and supervisory frameworks, water valuing, and incentives are mandatory to better allocate, control, and conserve water resources. Resource monitoring, decision-making under uncertainty, systems analyses, hydro-meteorological forecasting, and cautionary involvement of information systems. Savings in innovative technologies to improve productivity, keep and protect resources, reprocess stormwater and wastewater or groundwater, and develop alternative water resources ought to be explored to identify prospects to improve water storage. Ensuring swift dissemination and suitable adaptation or presentation of these advances will be critical to strengthening worldwide water security. Furthermore, innovative systems to improve water storage, including enrichment and reclamation of aquifers, should be explored to increase productivity, conserve and protect resources, reuse natural and wastewater, and take advantage of unconventional water sources. Rapid and appropriate adaptation of these advances is imperative to ensure global water security.

Healthy water management solutions for multifaceted water issues comprise unconventional knowledge and innovations integrated into water projects to intensify the impact. To this end, techniques operated in the spatial domain have the potential to successfully extract information about groundwater management and resources.

This book addresses the issues mentioned here in three sections:

1. "Introduction"

2. "Groundwater Management": This section provides an overview of the book, expounding governance and governability of groundwater in arid areas.

3. "Water Resources" discusses the identification of water sources in coal mines using machine learning modeling. It also reviews fractal scaling properties in rainfall time series.

This book asserts that platforms, sensors, data, techniques, and applications are all interconnected and work in tandem. While organizing the chapters according to their major goals, we also recognize that the authors of each chapter convey a complete story, at times ranging from the introduction of a method to a detailed application.

Bahareh Kalantar
RIKEN Center for Advanced Intelligence Project,
Goal-Oriented Technology Research Group,
Disaster Resilience Science Team,
Tokyo, Japan

Section 1

Introduction

Chapter 1

Characteristics and Assessment of Groundwater

Naseem Akhtar, Muhammad Izzuddin Syakir, Mohd Talha Anees, Abdul Qadir and Mohamad Shaiful Yusuff

Abstract

Groundwater system is very vital to humanity and the ecosystem. Aquifers are determined based on the absence or presence of water table positioning, that is, confined, unconfined, leaky aquifers and fractured aquifers. The objective of this chapter is to discuss the characteristic and assessment of groundwater within the scope of vertical distribution of GW, types of the aquifer system, types of SW-GW interface, and SW-GW interaction at both local and regional scales. The properties of the aquifer depend on the physical characteristics of the materials (porosity, permeability, specific yield, specific storage, and hydraulic conductivities) which are determined by techniques like resistivity surveys and pumping tests followed by remote sensing and geographic information system for better information on the groundwater system. Furthermore, understanding the SW-GW interactions through available methods (seepage meter, heat tracer, and environmental tracer) is useful in watershed management, that is, risk management and assessment of the aquifer system.

Keywords: aquifer characteristics, GW distribution, SW-GW interaction, SW-GW methods, resistivity survey, pumping test, RS, GIS

1. Introduction

Groundwater (GW) belongs to all subsurface water, including saturated and unsaturated zones. More than 1.5 billion inhabitants around the globe depend on the groundwater for agriculture usage and industrialization consumption. However, pollutions were identified as one of the major challenges in hampering GW withdrawal (**Figure 1**) [1]. The exchange of the chemical and physical characteristics in water will affect the quality of groundwater resources, hence leading to the availability of humans in terms of quantity [2].

Groundwater is deposited between the pore spaces of rock/soils, cracks, joints, and fractures and various geological formations. The movement of groundwater in soils and rocks depends on the hydraulic characteristics of the shape and size of void spaces. Water can flow easily through certain rocks through the soil into the underground aquifer system, but water typically penetrates through fractures, cracks, and some other geological formations. Generally, there are three distinct types of geological formations of groundwater that determine the availability of groundwater resource, namely, aquifers, aquitard, and aquiclude.

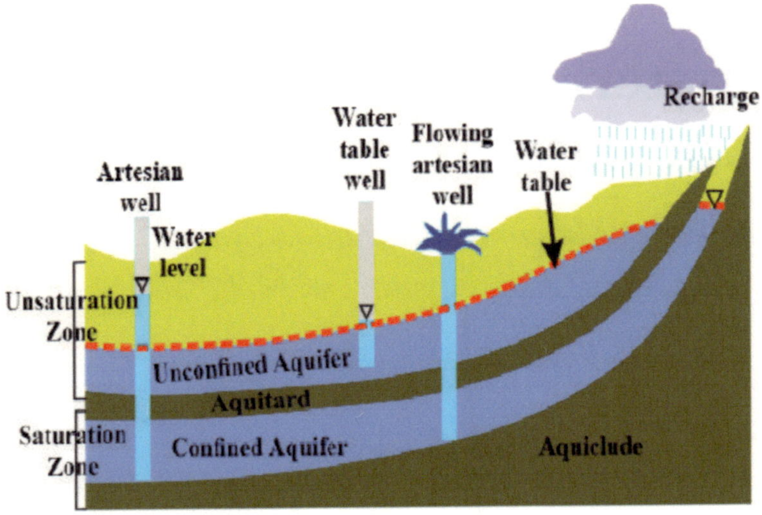

Figure 1.
The figure is showing unconfined aquifer and confined aquifer.

An aquifer is a highly permeable or porous saturated formation (conglomerate, sandstone, limestone, unconsolidated sand, gravels, fractured limestone, fractured basalt, etc.) that not only stores water but also provides adequate amounts of water and thus is considered as significant groundwater resources. The aquitard is a partially saturated formation (shale or clay) that allows water through it but does not provide enough available water than the aquifer. An aquiclude is an impermeable layer (clay) produces a considerable volume of water because of its high porosity but does not provide significant amount of water.

Groundwater passes from recharge zones to discharge zones along flow routes of variable lengths and comes in contact with surface water (SW) essentially at low elevated areas [3]. Surface water resources mostly depend on regional precipitation/rainfall and it may be lost by infiltration through the streambed, layer of soil-moisture, and cracks or fractures to interacts with the groundwater system and the area of mixing of both known as the hyporheic zone [4]. The interaction of surface water and groundwater takes several forms in which if surface water moves toward the groundwater system, it is referred to as a losing stream whiles the other way round is called gaining stream [5].

The surface water (SW)-groundwater (GW) interactions in the hyporheic zone take place within the close-streambed sediments at few scales, which depend on the hydraulic-potential strengths and bed geometry [6]. In SW-GW interactions and hyporheic exchange, earlier studies used three types of scales such as sediment scale (<1 m), local scale (1–1 km), and catchment scale (>1 km) [7]. However, Todd and Mays [8] classified only two scales such as local scale and regional scale which are associated with small watershed and large watershed, respectively. Here, SW-GW interaction is associated with the direction of streamflow, shallow GW aquifer property, and local GW flow system.

The scope of this chapter is to discuss the vertical distribution of GW, types of the aquifer system, types of SW-GW interface, and SW-GW interaction at both local and regional scales. This chapter has been divided into four sections; (i) groundwater distribution and aquifer characteristics, (ii) the SW-GW interactions at local and regional scale, (iii) types of SW-GW interface, and (iv) the methods for investigation of SW-GW interactions and aquifer system.

2. Groundwater distribution and aquifer characteristics

The distribution of groundwater is classified into two zones based on the water table, namely, unsaturation zone and saturation zone. The aquifer system is mainly divided into confined and unconfined aquifers, and its characteristics depend upon the main physical parameters such as porosity, permeability, transmissivity, specific yield, specific storage, and hydraulic conductivity.

2.1 Vertical distribution of groundwater

The groundwater occurrence is typically categorized into two major zones based on the water table namely unsaturation zone and saturation zone. The zone of unsaturation is also known as the zone of aeration (vadose zone), which is also subclassified into the soil moisture zone, intermediate vadose zone, and capillary zone. The unsaturation zone is comprised of interstices or void spaces that are partially filled with water and air. All interstices are fully saturated with water under hydrostatic pressure in the saturated zone under the water table.

2.1.1 Soil-moisture zone

The soil-moisture zone occurs across the main root zone beneath the earth's surface, but its thickness varies with the types of soil and vegetation. This zone plays a significant role in the recognition of hydrological processes [9] and also important for the interaction of the land-surface atmosphere [10]. The practices of agriculture and irrigation, particularly in arid and semiarid areas, primarily depend on the timely characterization of spatial and temporal soil moisture fluctuations in the root zone as a consequence of the soil moisture effect on health status and production of crops and salinization [11]. Several environmental factors such as physicochemical characteristics of water, surface slope and roughness, soil hydraulic conductivity, the porosity of the soil, and pre-existing soil pore moisture content are controlling the soil matrix's capacity to transfer water of which affecting the infiltration process [12].

2.1.2 Intermediate vadose zone

The intermediate vadose zone is located beneath the soil moisture zone and upper part of the capillary zone. Water that drops into this zone can be either drawn into the capillary interstices of the transition area through the molecular attraction or drawn downwards to the adjacent saturated zone.

2.1.3 Capillary zone

The zone is the lowest part of the aeration zone and directly above the water table where water as a component of the capillary action can be drawn back toward it. For a capillary zone of clay with a 0.0005 mm porous radius, the typical height may be 3 m, contrasted with fine sand of less than 10 cm with a 0.02 mm porous radius. Capillary water is the water stored above a surface of the water table in the capillary openings of unsaturated or saturated substances.

2.1.4 Saturation zone

The saturation zone is above the water table which is often referred to as the phreatic zone or aquifer system. Water that has profoundly infiltrated through the vadose zone enters the saturation zone and filled all pore spaces with water.

The thickness of the saturation zone varies from several meters beneath the earth's surface to numerous hundred meters. The factors to determine the thickness of this zone depend upon the local geology, accessibility of openings or pores in the rock formation, and water flow within the zone from recharging to discharge points. This saturation can take place range from several days or weeks to many months in duration. Moreover, groundwater is controlled by quantity and rainfall intensity, temperature, rock porosity, and permeability, dryness of the air, vaporization intensity during the rainy season, land slope, vegetative covering, and water absorption ability for soil. As well, significant volume of water can be contained within fractures and joints structures. The following are typical opening types contained in rock: (1) openings in gravel and sandstone formations with individual particles; (2) vugs, caverns, and solution channels in dolomite and limestone rock; and (3) joints, crevices, gas holes and faults in metamorphic rocks and igneous formations.

2.2 Types of aquifers

Aquifers are generally categorized into two major classifications, confined and unconfined aquifers; leaky and fractured aquifers are sometimes addressed in some other aquifers (**Figure 1**).

An unconfined aquifer is a layer of water-bearing formations or rocks that do not have a confining bed at the top of the groundwater which is referred to as the groundwater table where the pressure becomes equivalent to the atmospheric pressure. The variation of groundwater levels varies and depends on the pumping from the wells, permeability, area of recharge and discharge, in effect impacting the increasing or declining water rates in wells that are extracted from aquifers. The water table is free to rise or to fall which is often called the free or phreatic surface. Contour graphs and water table profiles of wells that use the water to determine water quantities available as well as water distributing, and movement may be prepared from elevations of wells. The perching water sources, as shown in (**Figure 1**) are a case of unconfined aquifers. Their high susceptibility to contamination is a major problem with non-confined aquifers. If something dumps on the surface, it will penetrate vertically and go down into the storage of groundwater.

2.2.1 Confined aquifer

The definition of the confined aquifer as "a formation in which the groundwater is isolated from the atmosphere at the point of discharge by impermeable geologic formations; confined groundwater is generally subject to a pressure greater than atmospheric" [13]. It is also known as "artesian or pressure aquifers" and it occurs mostly just above the base of confined rock bodies or layers which is mostly composed of clay that can protect it from surface pollution. Punctured wells from artesian aquifers are more prone to fluctuate with their depth of water because of changes in pressure than the amount in stored water. When such an aquifer is well penetrated, the water level should increase over the base of the confined layer, as illustrated by the flowing and artesian wells of (**Figure 1**). The water reaches a confined aquifer in a region in which the confining layer reaches the surface. The groundwater flow system into aquifers is frequently affected by gravity and geological formations in such areas either vertically or horizontally. A zone that provides water to a restricted area is considered a recharge area and water may even be leak into a restricted bed. Water ups and downs in confined aquifers penetrating wells mainly result from pressure changes instead of storage volume changes. Confined aquifers thus show only limited variations in storage and are predominantly used as conduits to move water to natural or artificial discharges from recharge areas.

2.2.2 Leaky aquifer

Aquifers that are fully unconfined or confined appear less often than aquifers that are leaky, or semi-confined. This is a common occurrence of plains, alluvial valleys, or former lake basins where a semi-pervious aquitard or semi-confining bed is underlain or overlain by a permeable layer. Pumping water from a well into a leaky aquifer eliminates water in two directions such as the vertical flow into the aquifer through the aquitard and the horizontal flow in the aquifer.

2.2.3 Fractured aquifer

The fractured rock aquifers vary from the subsurface water systems that are stored in the geological formation. Although sedimentary aquifers hold and move a significant amount of water between specific sedimentary granules through pore spaces, however, fractured rock aquifers hold and move water in an otherwise impermeable rock mass through as cracks, joints, and fractures (**Figure 2**).

Therefore, fractured rock aquifers have hydraulic characteristics that vary from those found in sedimentary aquifers with accessible water (common in terms to be described as bore yield) and are typically defined by nature (opening, size, and extent) and degree of interconnection between discontinuities in the rock mass. The long-term yield from well in fractured rock aquifers depends on the location of the degree of discontinuity and the relationship of discontinuities in the total mass of the rock instead of on the permeability of the geological substances near the extraction phase. The aquifers in fractured rock typically depend on the amount of precipitation that caused the surface water runoff of which considerably greater than in flat regions. Moreover, permeability fractured rock aquifers can also be dramatically decreased by the weight of the overlying rock mass as open spaces progressively decrease between fractures and cracks.

2.3 Characteristics of aquifer

Several properties that contribute to the identification and characterization of the aquifer are discussed.

2.3.1 Porosity

Porosity (n) is the intrinsic characteristic of a substance and refers to the amount of void or empty space in each material. The porosity (void space) occurs between

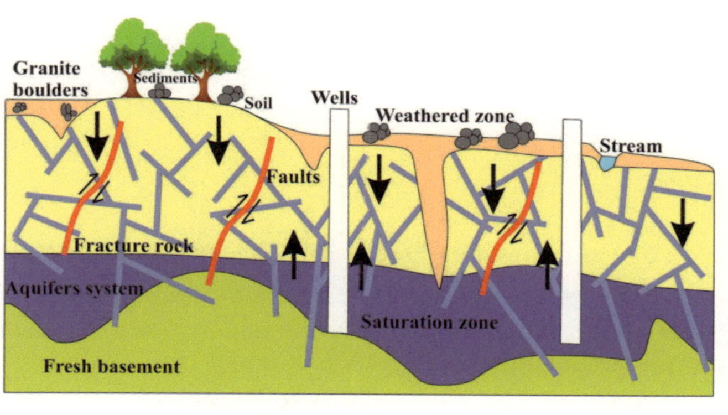

Figure 2.
The figure is showing the aquifer system in fractured rock formations.

the fragments of soil or rock. It is defined by the ratio between the volume of the void space and the volume of rocks/soils.

$$n = \frac{V_v}{V} * 100\% \tag{1}$$

where V_v is the volume of void space in a unit volume of earth material; and V is the unit volume of earth material (solids and voids).

2.3.2 Hydraulic conductivity and permeability

Permeability is defined as the ability of water movement through rock or soil which is directly related to porosity and it applies to the interconnected of pore spaces in rock or soil. Considering the relationship between driving and resisting forces on a microscopic scale during flow to porous media, hence, the permeability, k, is a function only of the area where the hydraulic conductivity K is defined:

$$k = \frac{K\mu}{\rho g} \tag{2}$$

where k is the permeability, K is the hydraulic conductivity, g is the acceleration due to gravity, ρ is the fluid density, and μ is the viscosity.

Hydraulic conductivity (K) is a physical characteristic that calculates the capacity of substance in the context of an applied hydraulic gradient to transfer water across the pore spaces and fractures of rock/soil [14]. It depends on various physical variables including porosity, the structure of the soil matrix, grain size distribution, type of soil fluid, particle arrangement, water contents, void ratio, and other factors [15, 16].

2.3.3 Transmissivity

The transmission (T) is the rate of discharge where the water is transferred under a hydraulic gradient over a unit width of an aquifer. It is calculated by a formula and expressed in m^2/s, or $m^3/day/m$ or l/day/m.

$$T = Kb(confinedaquifer) \tag{3}$$

$$T = Kh(unconfinedaquifer) \tag{4}$$

where K is the hydraulic conductivity, b is the aquifer thickness, and h is equivalent to the depth of confined aquifers.

2.3.4 Specific yield

Specific yield (S_y) as defined by Freeze and Cheery [14] is the storage term for unconfined aquifer where *the amount of water from the unconfined aquifer releases from the storage per unit surface area of aquifer per unit decline in the water table.* It is also known as unconfined storativity.

In other view, specific yield can be defined as the ratio of the volume of water that a saturated rock or soil will yield by gravity to the total volume of the rock or soil [15]. It is expressed in percentage.

$$S_y = \frac{V_w}{V} * 100\% \tag{5}$$

where V_w denotes the volume of water in a unit volume of earth materials; and V indicates the unit volume of earth material, including both voids and solids.

2.3.5 Specific storage

Specific storage (S_s) is the volume per unit amount of a saturated formation that is a deposit from the storage because of the compressibility of the mineral skeleton and the pore water per unit change in head. The specific storage is given by Jacob (1940) and is typically represented in cm^{-1} or m^{-1}.

$$S_s = \rho_{\omega} g(\alpha + n\beta) \tag{6}$$

where ρ_{ω} denotes water density, g is the acceleration of gravity, α shows compressibility of the aquifer skeleton, n indicate porosity, and β is the compressibility of water.

3. Groundwater and surface water interaction

Groundwater moves across flow paths arranged in space and develop a flow system. GW flow system is classified into local, intermediate, and regional flow systems (**Figure 3**) [17]. Water travels to the adjacent discharge area in a local flow system. One or more topographical low and high located between their discharge and recharge regions describe an intermediate flow system; however, contrary to the regional flux system, it does not occupy both the bottom of the basin and the major topographic high [18]. Water flows at a longer distance than the local flow system in a regional flow system and often discharge into large streams and lakes.

3.1 Characteristics of SW-GW interactions at the local scale

The range of groundwater at the local flow system is from 10 m to 10 km between the adjacent aquifers system and the stream reach. The recharge and discharge zones are associated with high and low areas respectively, associated with sub-watershed boundaries and local streams, respectively. The local GW flow system depends on the slope of topography and hydrogeology of the region (subsurface rock, streambed-sediment characteristics, and climatic conditions). At this scale, the seasonal effect on the hydrological response to recharge is high due to local flow systems, high water flux, and unsteady flow conditions. The fluctuation of the water table in local GW flow systems varies in different climatic conditions. For instance, the low water level in arid and semi-arid climate due to the low amount of precipitation and infiltration while the higher water level in a tropical environment is due to higher rainfall and infiltration. Therefore, SW-GW interaction is found more in a tropical and humid climate.

Generally, SW-GW can be introduced for homogeneous interaction of a stream and adjacent shallow aquifers system with hydrological processes, which is controlled by a SW-GW head and a streambed leakage coefficient. In hydrological processes, water moves with huge quantities of nutrients and streambed sediment and modifies the earth's surface through deposition and erosion. The hydrological processes give information about the drainage basin, small watershed, stream basin, evaporation, transpiration, evapotranspiration, runoff, and infiltrations rate (**Figure 4**). The hydrological exchange between GW and SW is through the downwelling and upwelling processes (**Figure 5**). Upwelling processes are those in which local GW flow moves toward SW and on the other hand, the situation is referred to as downwelling processes. During these processes, if the shape of the longitudinal streambed profile

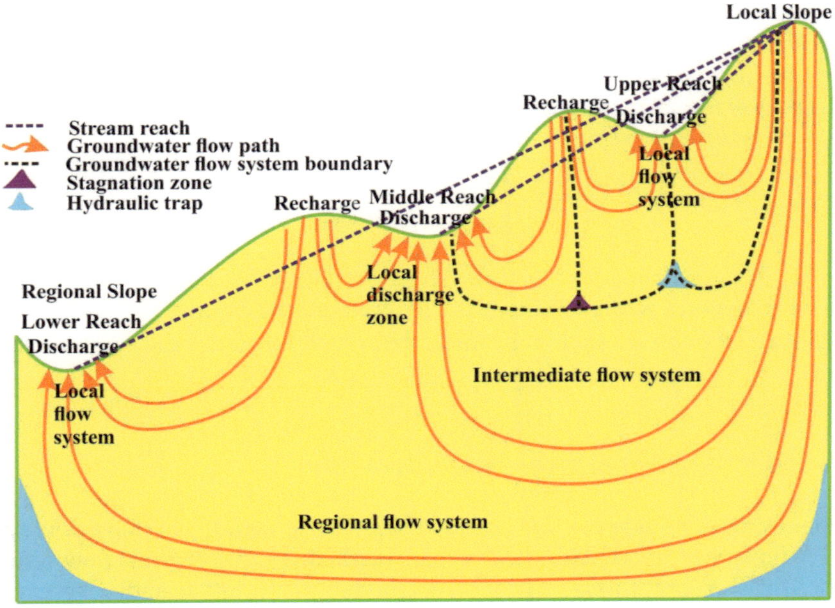

Figure 3.
GW flow system at the local and regional scale.

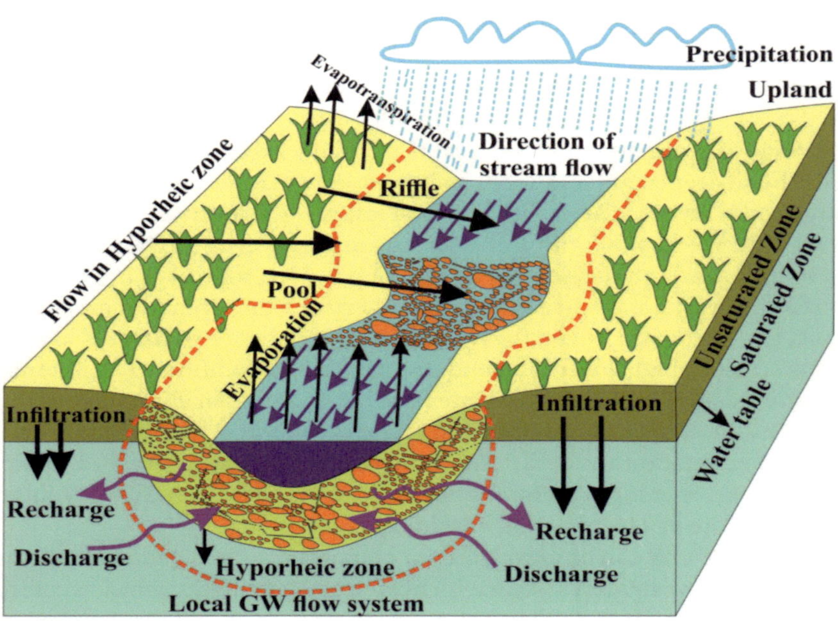

Figure 4.
River water and groundwater interactions in the hyporheic zone at the local scale.

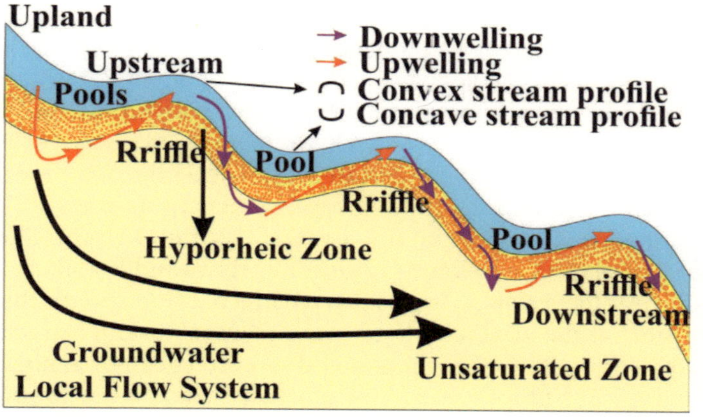

Figure 5.
Downwelling, upwelling, and hyporheic exchange processes.

is convex then the SW movement is through downwelling processes in the hyporheic zone whereas, if the shape of the longitudinal streambed profile of SW is concave then SW movement is through upwelling processes in the hyporheic zone [19]. The shapes of longitudinal streambed profiles are related to pool-riffle sequence and sediment bars, dunes, and ripples. The movement of stream water from riffles to pools is showing in (**Figure** 5) which is affected by the channel's sinuosity and bed load materials.

3.2 Characteristics of SW-GW interactions at the regional scale

The interaction of SW-GW is related to low topographical pathways varying from 10 to 100 km or more at a regional scale. The recharge or discharge

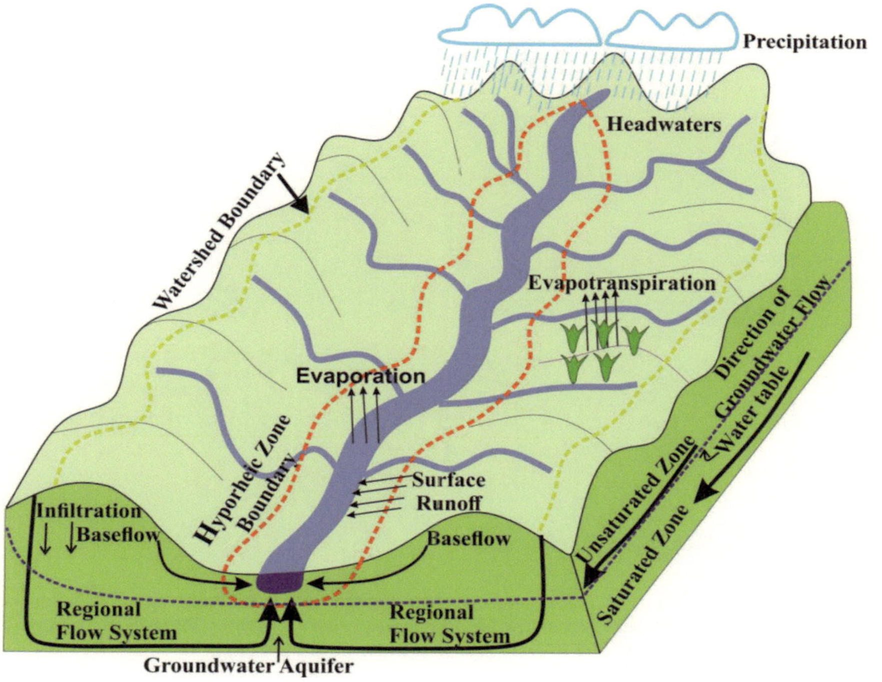

Figure 6.
River water and groundwater interactions at the regional scale.

trends of groundwater, regional topographic, hydrological conditions are mainly characterized by the regional groundwater flow system at a regional scale. The hydrological processes of the large watershed such as precipitation, surface run-off, infiltration, evapotranspiration, base flow, streamflow, and channel conditions are described in **Figure 6**. All hydrological processes can cover on the regional scale of a large watershed and a small stream reach conditions can cover local scales. Slow hydrological response to recharging areas will result in an insignificant seasonal impact on the regional GW flow system. The regional flow system is developed with long-distance SW flows and low charging levels on higher topographical slopes. It explains why recharging and discharge levels fluctuate more at the local level than at the regional level. The recharge or discharge rate investigation of the regional flow system can be analyzed using environmental tracer isotopes and hydro-chemical characteristics (major ions and heavy metals) [6].

4. Different types of SW-GW interactions

There are several types of interactions between GW and SW. A losing river does not lose water as it flows downstream by percolation, but it can also lose water through evaporation, use of plants, and consumption of human activities.

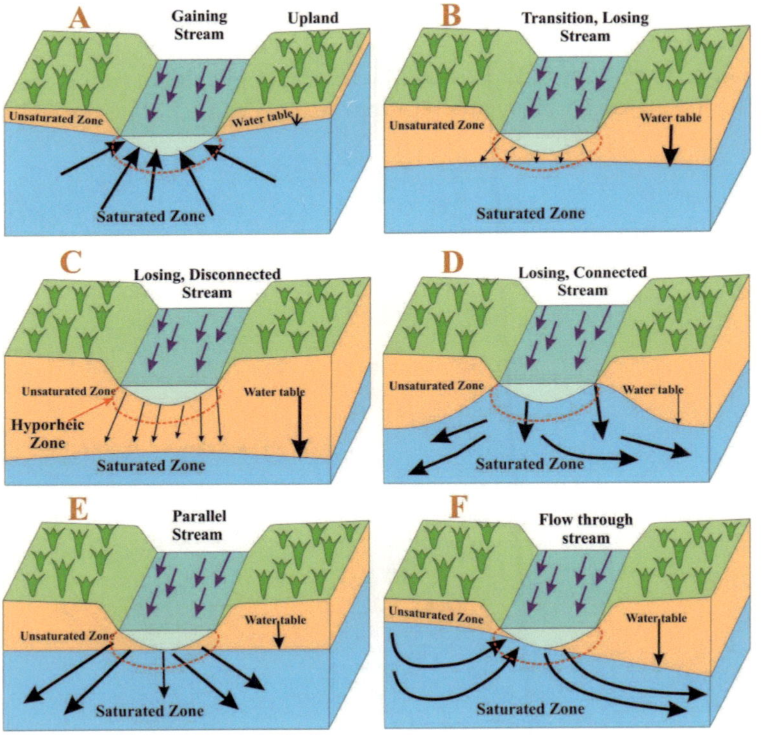

Figure 7.
Types of stream water and GW interactions, while (A) a stream water gains from GW, known as gaining stream, (B) a stream loses water to GW at shallow depth known as disconnected or losing or transition stream, (C) a stream loses water to GW at great depth known as a disconnected stream, (D) a stream loses water, but connected with GW known as a connected stream (E) a stream stage and the GW head are equally known as the parallel Stream, and (F) a stream shows the gain on one side and the loss on the other side to GW known as flow-through stream.

4.1 Gaining stream or effluent stream

In this connection, the level of GW is higher than the riverbed which recharges the river (**Figure 7A**). It can also be described as entering of GW into SW when SW reaches its base level which results in gaining stream connection.

4.2 Losing stream or influent stream

In this connection, given the level of GW is lower than the streambed, hence, SW recharges the GW. Losing streams connection divided into two types in which it is either connected or disconnected with the GW table. The term "transition" is used to define the condition of connected and disconnected streams. As shown in (**Figure 7B**), the unsaturation zone is presented in transition with a shallow GW table between riverbed. Note, there are distinguished interactions of the disconnected stream with shallow and deep water table [20].

4.3 Losing disconnected stream

In this connection, the unsaturation zone of sediments exists between the channel and regional water table hence it can be said that the system may be hydraulically disconnected (**Figure 7C**). The term disconnected has been criticized because it can suggest a system where there is no exchange of recharge and discharge of the GW system [21]. Therefore, the rate of infiltration of a disconnected system has been referred to as a "maximum losing condition"—stream discharge mechanism. Thus, the water table occurs at greater depth in the disconnected system and at shallow depth in the transition zone.

4.4 Losing connected stream

A stream is a stream that loses water while it flows downstream. The water penetrates the ground and recharges the local GW flow as the water table lies below the level of the channel with the absence of an unsaturation zone (**Figure 7D**).

4.5 Parallel stream

This interaction occurs when the stage of the stream and the head of the groundwater is equal (**Figure 7E**).

4.6 Flow-through stream

Where the stage of the channel is less than the head in the groundwater on one bank side and larger than the head in the groundwater on the opposite bank side, this process is seen as a flow-through reaches (**Figure 7F**). This interaction occurs most frequently when the stream cuts perpendicular to the regional GW flow, which in the case of fluvial plains is along their axis.

5. Methods

The methods for the investigation of aquifer systems such as remote sensing (RS) and geographic information system (GIS), resistivity test, and pumping tests will be discussed. Several related approaches also will be discussed such as seepage meter, Darcy's law, heat tracer method, and environmental tracer method for the investigation of SW-GW interactions.

5.1 Resistivity survey

In groundwater system, evaluation geophysical methods (geothermal gravity, electrical resistivity, etc.) have been well recognized. The electrical resistivity survey is one of the tools that is very effective to identify subsurface profiles without interfering with the structure of the soil [22]. The usage of this method enables the measurement of groundwater quantities and quality. This includes detailed knowledge concerning the geological and hydrological information of the GW system such as subsurface mapping to identify aquifer-protective structures, the analysis of infiltration of the vadose zone, measuring the extent of volume and internal aquifer structure, and groundwater contamination [23]. It is effectively used to estimate soil porosity and soil permeability as a non-destructive process. In addition, it is commonly utilized for the interaction of changes in the resistivity of the subsurface with the soil characteristics. The negligible porosity and permeability of the hard rock, as well as igneous and metamorphic rocks, in terms of soil exploitability, but the alteration processes taking place in the first 10–100 m of depth can significantly increase their fracture permeability. In the zone influenced by modification, this may create moderate secondary porosity aquifers.

These aquifers are very critical for irrigation and the availability of potable water in many parts of the world. Altered methods often influence the overall porosity of the rock such as water content which results in a varying spectrum of electric resistivity within the transition region [24]. Consequently, it is a good potential technique for the study of alteration zones in hard rocks, electrical resistivity in rocks influenced by differing weathering degrees. Schlumberger array system [25] was used to perform the resistivity survey. "ABEM SAS 1000 Terrameter" was the device used for performing Vertical Electrical Sounding (VES). For resistance measurement, four electrodes were selected at a certain time. Two existing electrodes situated on the outside of the potential electrodes were inserting currents into the field. The potentially different electrodes were quantified and the ground resistance was measured by Eq. (2).

$$\rho = K_g * resistanceofearth = K_g * \left(\frac{V}{I}\right) \tag{7}$$

$$K_g = \left\{\pi\left(\frac{AB}{2}\right)^2 * \left(\frac{MN}{2}\right)^2\right\} / MN \tag{8}$$

where ρ denotes the apparent resistivity (ohm-meter), K_g is the geometric constant, I indicate current (ampere), V is the voltage (volt), AB is spacing between the current electrode (m) and MN is spacing between the potential electrode (m). The geometric factor (K_g) is based on field observation calculation and by multiplying the geometric factor with data of resistivity the apparent resistivity values can be calculated. For instance, the transmissivity estimates were 0.588, 0.578, and 0.756 m^2/min, respectively, by the analysis of grain size distribution, the resistivity survey, and the pumping test [25]. The finding on the results indicated that the values of aquifer transmissivity have been found much similar to each other by measurement of grain size distribution, pumping test, and resistivity survey.

5.2 Pumping test

Pumping test is a field technique and it is used for the assessment of the aquifer characteristics such as hydraulic conductivity (*K*), storage coefficient (*S*), and transmissivity (*T*). Aquifer hydraulic parameters are spatially and temporarily influenced by their heterogeneity, complicated geologic conditions, as well as multi-part boundaries but these characteristics in various aquifer areas, are challenging to describe efficiently [26]. The geological formation of the aquifer (confined, leaky, unconfined, and fractured aquifer) influences the hydraulic parameters to estimate; thus, various interpretive techniques are applied. Implementation of geophysical studies and pumping test techniques may be used to maximize the comprehension of hydrogeology models by accurately detecting such essential aquifer characteristics: permeability, thickness, porosity, transmissivity, hydraulic conductivity, etc. Various pumping test methods are used to determine aquifer hydraulic characteristics; but, long term, step pumping, and recovery tests are mostly utilized. Aquifer characteristics can be found by using easy methods such as the first analytical solutions proposed by [27] Thiem (1906) for a steady-state condition that gives an equation for the groundwater flow in aquifers subject to pumping. After this, [28] Theis (1935) and Cooper and Jacob (1946) [29] find extremely restrictive conditions in terms of a transient state that limit their implementation to aquifers that are uniform, homogeneous and isotropic, constant thickness, porous and permeable which produce pumping with a constant discharge in a completely penetrating well. The following formulas calculate the aquifer properties by Theis-Jacob method:

$$s(r,t) = \frac{Q}{4\pi T}\left[W\left(\frac{4^2 S}{4Tt_p}\right)\right] \tag{9}$$

$$T = KB = \frac{\gamma b_m^3}{12\mu} N_f \tag{10}$$

where Eq. (5) shows that the $s(r,t)$ is the drawdown, Q is the pumping rate, T is the hydraulic transmissivity, W is the well function for a confined aquifer, S is the aquifer storativity and t_p is the pumping period (t) for $t \le t_p$. The Eq. (6) is showing that K is the hydraulic conductivity, N_f is defined by imposing $N_f \times b_m^3 = n_e \times B$, and n_e is the effective porosity ($\cong 0.003 \pm 0.002$) of the studied fractured aquifer given by tracer tests [29]. For example, Alfy et al. [30] were used the pumping test and geophysical logging for the investigation of the hydraulic and petro-physical characteristics of the folded UmmerRadhuma (UeR) Formation, Saudi Arabia. The findings were obtained showing that, concerning efficient porosity, permeability, hydraulic conductivity, and transmissibility, the average values of 220%, >100 mD, 3.30×10^{-5}–1.34×10^{-3} m/s, and 1.49×10^{-3}–6.04×10^{-2} m²/s.

5.3 Remote sensing (RS) and geographic information system (GIS)

Intensive performance applications of RS and GIS are spatial data analysis and monitoring methods for groundwater sources. RS data integration with the GIS environment seems to be very beneficial in considerably identifying the specific groundwater potential areas. In the short time available, RS and GIS cover a vast and unacceptable region of the earth's surface to assess areas of possible groundwater and to identify natural recharging locations [31]. RS and GIS information are valuable for many geological resources including mineral exploration, hydrogeology conditions, structural, geomorphological, lithological features, depth, and thickness of the aquifer system and other geological areas [32]. Furthermore, in the area of groundwater studies, researchers have used thematic layers such as geomorphology, geology, drainage patterns, lineaments, vegetation, intensities of rainfall, and slopes [33]. A geophysical resistivity survey was performed by [34] and borehole lithology results were compared for aquifer characterization with groundwater potential mapping which was created by RS and GIS. For a hydrogeological study, [35] integrated electrical resistivity survey data with RS outputs in a GIS environment. Moreover, [36] suggested that the geophysical resistivity data integrated with high-resolution satellite data collected from RS and GIS techniques provide more accurate information on geological and hydrological characteristics and also give possible groundwater potential zones in the hard rock formations.

5.4 Seepage meter

Seepage meter is one of the most common instruments for directly measuring SW-GW seepage flux. Initially, it was developed to measure water loss from a canal in 1940 by [37], and also it is used for other purposes such as measuring seepage flux in small lakes, estuaries, rivers, and several other environments [38]. The basic concept of the seepage meter is the difference between initial (V_o) and final (V_f) volume of water through a surface area (A) in time (t) and is given as

$$Seepageflux\,(Q) = \frac{\left(V_f - V_o\right)}{tA} \tag{11}$$

This method was described as a plastic bag type seepage meter, which is based on isolating principle and covers a portion of SW-GW interactions with a bottomless cylinder which important in determining the directions of water exchange at the local scale [38]. The streambed features (riffle-pool sequences) can be recognized by seepage meters at the local scale because this method is useful to investigate the water flux estimates of lower streambed sediments [39]. Seepage meters can also be used to determine the volumetric change in flow, discharge, or recharge zones along with the streambed sediments in the hyporheic zone at a local scale. The seepage meter is favorable for those streams which have low current velocities which represent a local scale stream [17]. For instance, four seepage meters have been used, along with riverbed sediments of Biebrza River, Poland to quantify the hyporheic exchange flux at a local scale [40].

5.5 Heat tracer method

Subsurface temperature variation is associated with the movement of water. This variation affects the chemistry of water which can be traced by the heat tracer method. According to Anderson [41], the heat tracer method is used to determine hyporheic exchange, GW flow patterns, and rate of discharge and recharge at the local regional scale. SW temperature fluctuates throughout the season and also daily while GW temperature remains constant throughout the year. This method has been used by Schmidt, Raich, and Schirmer [42] for SW-GW interactions at the local scale and suggested that streambed temperatures can be quickly, reliably, and cheaply assessed the SW-GW interactions at several locations. The successful combination of their conceptual methods described by Constantz [43] with these technical improvements to assess SW-GW interactions, GW discharge or recharge, SW movement through the streambed, and GW flow systems. In the past, heat tracers' methods have been used to evaluate losing and gaining stream. For instance, the temperature was investigated by Cox, Su, and Constantz [44] and also determined the special conductance, and chloride between the aquifers system using the heat tracer method in the Russian River, Mendocino, California. Their results indicated that the special conductance and chloride data were not correlated with RW data. It means GW was not significantly varied by the exchange of SW and GW system and temperature variations in GW were negligible.

5.6 Darcy's law method

Darcy's law [45] measures the hydraulic gradients, aquifer hydraulic conductivity, cross-sectional area of the aquifer perpendicular to the flow, and to evaluate the rate of GW flow. Darcy's law expressed as,

$$\text{Darcy's } law\,(q) = -K\frac{(dh)}{dl} \tag{12}$$

where q is a specific discharge (L/T), K is hydraulic conductivity, l is the distance (L), and h is the hydraulic head. The hydraulic gradient is measured by piezometers and mini-piezometer at both local and regional scales. Piezometer indicates that the hydraulic head difference at great depth or vertical GW flow while mini-piezometer at shallow depths indicate GW downwelling or upwelling processes. The hydraulic conductivity is based on streambed sediments. It can be utilized as the estimation of streambed sediments by the slug test. A slug test is based upon the immediate increase or fall of the water level in the bore and the conformity of the water level to the original position when the water returns. The velocity and direction of GW flow can be determined by the mini-piezometer method. Furthermore, the estimation of flow between the SW-GW aquifer systems through semi-impervious stratum in one dimension which is used based on Darcy's law.

$$\text{Darcy's } law\,(q) = K\Delta h \tag{13}$$

where q flows between SW-GW, Δh is the river head and aquifer head, and K is hydraulic conductivity of the semi-impervious streambed stratum. The investigation of the Platte River by Chen et al. [45] in eight tributaries of eastern and south-central Nebraska, the USA, with the help of this method. The river joins the Missouri River in the eastern part of Nebraska and its interactions with high plains aquifer systems.

5.7 Environmental tracer method

Environmental tracer method is used to analyze the SW-GW interactions on both local and regional scales which are based on isotope data and geochemical data such as major ions or heavy metals. Stable hydrogen and oxygen isotopes are useful for assessing the flow of precipitation, source of water, age of water, and hydrological processes. In addition, this method has been used to determine the GW influx to a tropical river with major ions to supplement GW flux results [46]. Furthermore, it has been used to determine the gaining, losing disconnected and connected stream reach condition based on the geochemical parameters and stable isotopes [47–53]. Moreover, it is important to understand the recharge or discharge zone along with the GW flow system.

6. Conclusions

The number of groundwater studies continues to increase globally. This chapter discusses the SW-GW interactions and groundwater characteristic with appropriate assessment methodologies encompassing subsurface investigation (resistivity), hydraulics aspects of groundwater (pumping test), and mapping (RS and GIS).

Furthermore, understanding the SW-GW interactions through available methods (seepage meter, heat tracer, and environmental tracer) is useful in watershed management, that is, risk management and assessment of the aquifer system. Moreover, environmental tracer method is also a useful for the evaluation of the hydrological process, source of water, age of water, and gaining and losing disconnected and connected stream conditions.

Conflict of interest

The authors declare no conflict of interest.

Author details

Naseem Akhtar*, Muhammad Izzuddin Syakir, Mohd Talha Anees, Abdul Qadir and Mohamad Shaiful Yusuff
School of Industrial Technology, University Sains Malaysia, Pulau Pinang, Malaysia

*Address all correspondence to: naseemamu6@gmail.com

References

[1] Akhtar N, Rai SP. Heavy metals concentrations in drinking water and their effect on public health around moth block of Jhansi District, Uttar Pradesh, India. Indian Journal of Environmental Protection. 2019;**39**(9):945-953

[2] Fleckenstein JH, Krause S, Hannah DM, Boano F. Groundwater-surface water interactions: New methods and models to improve understanding of processes and dynamics. Advances in Water Resources. 2010;**33**(11):1291-1295

[3] Conant B, Robinson CE, Hinton MJ, Russell HAJ. A framework for conceptualizing groundwater-surface water interactions and identifying potential impacts on water quality, water quantity, and ecosystems. Journal of Hydrology. 2019;**574**:609-627

[4] White DS. Perspectives on defining and delineating hyporheic zones. Journal of the North American Benthological Society. 2006;**12**(1):61-69

[5] Brunner P, Simmons CT, Cook PG. Spatial and temporal aspects of the transition from connection to disconnection between rivers, lakes and groundwater. Journal of Hydrology. 2009;**376**(2):159-169

[6] Barthel R, Banzhaf S. Groundwater and surface water interaction at the regional-scale – A review with focus on regional integrated models. Water Resources Management. 2016;**30**(1):1-32

[7] Dahl M, Nilsson B, Langhoff JH, Refsgaard JC. Review of classification systems and new multi-scale typology of groundwater-surface water interaction. Journal of Hydrology. 2007;**344**(1-2):1-16

[8] Todd DK, Mays LW. Grounwater Hydrology. 3rd ed. USA: Hamilton Printing; 2005

[9] Western AW, Zhou SL, Grayson RB, McMahon TA, Blöschl G, Wilson DJ. Spatial correlation of soil moisture in small catchments and its relationship to dominant spatial hydrological processes. Journal of Hydrology. 2004;**286**(1-4):113-134

[10] Western AW, Blöschl G. On the spatial scaling of soil moisture. Journal of Hydrology. 1999;**217**(3-4):203-224

[11] Mahmood R. Scale issues in soil moisture modelling: Problems and prospects. Progress in Physical Geography. 1996;**20**(3):273-291

[12] Hupet F, Lambot S, Javaux M, Vanclooster M. On the identification of macroscopic root water uptake parameters from soil water content observations. Water Resources Research. 2002;**38**(12):36-1-36-14

[13] Driscoll FG. Groundwater and wells. 2nd ed. St. Paul, Minnesota: Johnson Filtration Systems; 1986

[14] Freeze RA, Cherry JA. Groundwater. Bergen Country, NJ, USA: Prentice Hall Inc; 1979

[15] Kalhor K, Ghasemizadeh R, Rajic L, Alshawabkeh A. Assessment of groundwater quality and remediation in karst aquifers: A review. HHS Public Access. 2019;**8**:104-121

[16] Johnson AI. Specific yield compilation of specific yields for various materials—Hydrologic properties of earth materials. Geological Survey Water-Supply Paper. 1967;**1662-D**:80

[17] Tóth J. A theoretical analysis of groundwater flow in small drainage basins. Journal of Geophysical Research. 1963;**68**(16):4795-4812

[18] Sophocleous M. Interactions between groundwater and surface

water: The state of the science. Hydrogeology Journal. 2002;**10**(1):52-67

[19] Harish HK, Arina K. Ground water and surface water interaction. In: GIS and Geostatistical Techniques for Groundwater Science, Section C., G. Q. A. U. G. and G. A. 14.2. Elsevier Inc; 2019. pp. 197-207

[20] Peterson DM, Wilson JL. Variably Saturated Flow Between Streams and Aquifers. Socorro: New Mexico Water Resources Research Institute; 1988

[21] Brunner P, Cook PG, Simmons CT. Connected and disconnected streams. Water Resources Research. 2010;**66**:37-39

[22] Larsen F et al. Controlling geological and hydrogeological processes in an arsenic contaminated aquifer on the Red River flood plain, Vietnam. Applied Geochemistry. 2008;**23**(11):3099-3115

[23] Almadani S, Ibrahim E, Al-Amri A, Fnais M, Abdelrahman K. Delineation of a fractured granite aquifer in the Alwadeen area, Southwest Saudi Arabia using a geoelectrical resistivity survey. Arabian Journal of Geosciences. 2019;**12**(15)

[24] Gowda JP, Kumar SCP. Segmentation of white blood cell using k-means and Gram-Schmidt Orthogonalization. Indian Journal of Science and Technology. 2017;**10**(6):1-5

[25] Keller GV. Dipole method for deep resistivity studles. Geophysics. 1966;**31**(6):1088-1104

[26] Arshad M, Shakoor A, Ahmad I, Ahmad M. Hydraulic transmissivity determination for the groundwater exploration using vertical electric sounding method in comparison to the traditional methods. Pakistan Journal of Agricultural Sciences. 2013;**50**(3):487-492

[27] Calvache ML, Sánchez-Úbeda JP, Duque C, López-Chicano M, De La Torre B. Evaluation of analytical methods to study aquifer properties with pumping tests in coastal aquifers with numerical modelling (Motril-salobreña aquifer). Water Resources Management. 2015;**30**(2):559-575

[28] Thiem G. Hydrologische Methoden. Leipzig, Germany: JM Gebhardt; 1906a

[29] Theis CV. The relation between the lowering of the piezometric surface and the rate and duration of discharge of a well using ground-water storage. Eos Transactions American Geophysical Union. 1935;**16**(2):519-524

[30] Cooper HHJ, Jacob CE. A generalized graphical method for evaluating formation constants and summarizing well-field history. Eos Transactions American Geophysical Union. 1946;**27**(4):526-534

[31] Caputo MC, Carlo DL, Masciale R, Masciopinto C. Long-term pumping test and ert to visualize hydrogeologic barriers in heterogeneous and Karstic coastal aquifers. Journal of Geology and Geophysics. 2017;**6**(5):1-5

[32] Alfy ME, Lashin A, Faraj T, Alataway A, Tarawneh Q , Al-Bassam A. Quantitative hydro-geophysical analysis of a complex structural karst aquifer in Eastern Saudi Arabia. Scientific Reports. 2019;**9**(1):1-18

[33] Senthilkumar M, Gnanasundar D, Arumugam R. Identifying groundwater recharge zones using remote sensing & GIS techniques in Amaravathi aquifer system, Tamil Nadu, South India. Sustainable Environment Research. 2019;**29**(1):1-9

[34] Elubid BA et al. Geospatial distributions of groundwater quality in Gedaref state using geographic information system (GIS) and

drinking water quality index (DWQI). International Journal of Environmental Research and Public Health. 2019;**16**(5):1-20

[35] Anbazhagan S, Jothibasu A. Geoinformatics in groundwater potential mapping and sustainable development: a case study from southern India. Hydrological Sciences Journal. 2016;**61**(6):1109-1123

[36] Oseji JO, Atakpo EA, Okolie EC. Geoelectric investigation of the aquifer characteristics andgroundwater potential in Kwale, Delta state, Nigeria. Journal of Applied Sciences and Environmental Management. 2005;**9**(1):157-160

[37] Kumar MG, Bali R, Agarwal AK. Integration of remote sensing and electrical sounding data for hydrogeological exploration—A case study of Bakhar watershed, India. Hydrological Sciences Journal. 2009;**54**(5):949-960

[38] Rashid M, Lone MA, Ahmed S. Integrating geospatial and ground geophysical information as guidelines for groundwater potential zones in hard rock terrains of south India. Environmental Monitoring and Assessment. 2012;**184**(8):4829-4839

[39] Israelson OW, Reeve RC. Bulletin No. 313—Canal Lining Experiments in the Delta Area, Utah, UAES Bulletins; 1944

[40] Brodie RS, Baskaran S, Ransley T, Spring J. Seepage meter progressing a simple method of directly measuring water flow between surface water and groundwater systems. Australian Journal of Earth Sciences. 2014;**42**(14):37-41

[41] Lee DR. A device for measuring seepage flux in lakes and estuaries. Limnology and Oceanography. 1977;**22**(1):140-147

[42] Kalbus E, Reinstorf F, Schirmer M. Measuring methods for groundwater-surface water interactions: A review. Hydrology and Earth System Sciences. 2006;**10**(6):873-887

[43] Anibas C, Buis K, Verhoeven R, Meire P, Batelaan O. A simple thermal mapping method for seasonal spatial patterns of groundwater-surface water interaction. Journal of Hydrology. 2011;**397**(1-2):93-104

[44] Anderson MP. Heat as a ground water tracer. Ground Water. 2005;**43**(6):951-968

[45] Schmidt C, Raich MB, Schirmer M. Characterization of spatial heterogeneity of groundwater-stream water interactions using multiple depth streambed temperature measurements at the reach scale. Hydrology and Earth System Sciences. 2006;**10**(3):1419-1446

[46] Constantz J. Heat as a tracer to determine streambed water exchanges. Water Resources Research. 2008;**46**(4):1-20

[47] Cox MH, Su GW, Constantz J. Heat, chloride, and specific conductance as ground water tracers near streams. Ground Water. 2007;**45**(2):187-195

[48] Darcy H. Les fontaines publiques de la ville de Dijon. Paris: Victor Dalmont; 1856

[49] Chen X, Dong W, Ou G, Wang Z, Liu C. Gaining and losing stream reaches have opposite hydraulic conductivity distribution patterns. Hydrology and Earth System Sciences. 2013;**17**(7):2569-2579

[50] Cook PG. A Guide To Regional Flow in Fractured Rock Aquifers. CSIRO Land and Water, Glen Osmond: Henley Beach, South Australia: Seaview Press; 2003

[51] Engelhardt I et al. Comparison of tracer methods to quantify

hydrodynamic exchange within the hyporheic zone. Journal of Hydrology. 2011;**400**(1-2):255-266

[52] Xu W, Su X, Dai Z, Yang F, Zhu P, Huang Y. Investigations multitraceurs des interactions nappes-rivières: cas d'étude du bassin de la rivière Nalenggele, nord-ouest de la Chine. Hydrogeology Journal. 2017;**25**(7):2015-2029

[53] Aguilar BJ, Banks EW, Batelaan O, Kipfer R, Brennwald MS, Cook PG. Groundwater residence time and aquifer recharge in multilayered, semi-confined and faulted aquifer systems using environmental tracers. Journal of Hydrology. 2017;**546**:150-165

Section 2

Groundwater Management

Chapter 2

Governance and Governability of Groundwater in Arid Areas

Edwin Pino and André Steenken

Abstract

The crisis of governance and governability in the use of groundwater in coastal aquifers in arid zones is an element that contributes to the depletion and deterioration of the groundwater quality, due to processes of marine intrusion. Under the current exploration conditions, the governance crisis is characterized in the overall consumption of groundwater by humans, their animals, and by their industrials. Interpreting current national and international regulations applicable to this type of system confirms this area of investigation. This problem goes beyond the laws in all its terms regarding the management regimes and limitations on the availability of groundwater resources. If immediate actions are not taken by the government, the crisis of the system could become irreversible, with the consequent economic damage that this involves.

Keywords: governance, governability, depletion and deterioration of groundwater, arid areas

1. Introduction

The arid regions have, as their main limitation for their development, the serious shortage of water. That is the reason why established producers in these areas use groundwater, this being probably the only source of supply. Currently, most of the coastal aquifers are being over-exploited, which results in a gradual and permanent decrease in the water table, compromising their non-renewable reserves, causing the phenomenon of marine intrusion. Faced with this problem, government entities have been preparing plans for the exploitation of groundwater, with the purpose of exploiting the water resource in an efficient, organized, and sustained manner. In the same way, to evaluate the hydrogeological conditions and characteristics of the subsoil of the aquifer, which allows determining the exploitable reserve of water in quantity and quality, feasible to exploit, in a long-term sustainable way, and propose measures to improve its management and conservation.

In these aquifers, the water shortage has been exacerbated in recent decades due to the frequent presence of droughts, the expansion of the unauthorized agricultural frontier with the consequent incorporation of new wells without a license to exploit groundwater, the accelerated population growth of cities, and the multiple economic activities that this brings with it. However, these aquifers, due to the lack of an adequate exploitation program, have been overexploited, which has resulted in a gradual and permanent decrease in the water table, compromising their non-renewable reserves, which has caused the phenomenon of marine intrusion due to the high concentration of underground water exploitation wells in the area close to the sea line. Rapid economic expansion creates serious problems with the use of groundwater in arid zones, which normally have high rates of depletion [1].

The world's main aquifers on which hundreds of millions of people depend are depleting at an alarming rate according to data obtained through Grace satellites and released by NASA of the 37 largest aquifers scattered worldwide, from India and China to the USA or France, 21 of them have exceeded their sustainability point, which means that more water has been extracted from them than has been incorporated over the 10 years of observation [2]. The NASA study confirms the suspicions that many researchers already had, especially in cases of aquifers that are not rechargeable due to rain. Likewise, a concern in the field of environmental sciences and environmental law itself is the notable deterioration and decrease of wetlands and groundwater, taking into account that one of its causes is the phenomenon of climate change [3].

Along with the natural processes that generate water deficit problems in aquifer systems, there are also problems of a social nature. Governance occurs in the sense that the State is considered as the central agent of society's leadership; thus, their concern centers on "the ability to govern," considering society as the entity to be governed and administered [4]. Thus, from the perspective of governance, the problem and its solution emanate from the capacities of the government, with some independence from society, while Governance is part of the recognition that the government lacks sufficient capacities to solve social problems.

Therefore, the objectives of this chapter are as follows: (a) we will refer to governability when it comes to the institutional part conferred on the government and its institutions, that is, its capacity and range of action; (b) we will use governance when we refer to the joint action of government and society for a common positive objective (such as development), in order to achieve a balance. (c) In order to achieve integrated water management goals, it is necessary to harmonize the conditions and environment dynamics of the populations in relation to the hydrographic basins and the hydrological cycle [5].

2. Governance and governability

Governance is denoted to have two closely related faces: the capacity to formulate adequate policies and the capacity to implement them. These capacities require the construction of consensus, the assembly of management systems, laws, national and regional public agencies, knowledge and information, practices, and the adequate administration of the system that implies social participation and the development of competences [6]. In the specific case of Chile, it has a water management system (SGA-CH) that can be considered as the only one of its kind, it considers water as a tradable commodity. This system is legally based on the National Water Code (CNA), which is at the same hierarchical level as the civil code [7].

After an exhaustive bibliographic review, seven versions of governance are identified that are highly relevant due to their degree of expression in the Colombian context and that are applicable to the South American sphere. These interpretations are shown in **Table 1** and **Figure 1** [8].

If we want to make optimal use of our valuable groundwater resources and ensure their sustainability, we must manage them with care [9]. In practice, the hierarchical and centralized management scheme in force for more than four

Governance approaches	Definition
Good government	It refers to processes and institutions that deliver results that satisfy the needs of society by making good use of available resources
Governing without government	Uses the laws of nature instead of legislation and one-way communication instead of dialog
Networks	Consists of several elements that are connected to each other
Self-government	Refers to the control of an organization by people independent of the central or local government
Economic governance	Refers to the policy and regulatory settings that governments adopt to manage the economy
Corporate governance	It refers to the set of rules, practices and processes through which a productive activity is directed and controlled. Corporate governance essentially involves a balancing of the interests of users, the government and the community
New public management	Is an approach to public service organizations that is used in government and public service institutions and agencies, at both sub-national and national levels

Table 1.
Governance approaches.

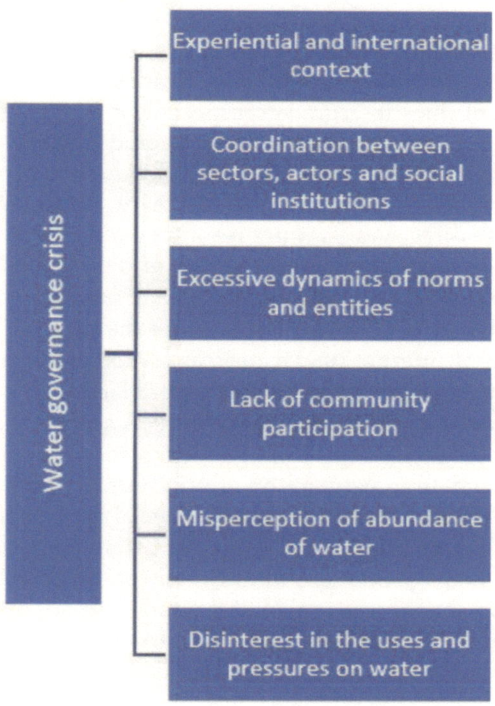

Figure 1.
Elements about the water governance crisis.

decades has been continued, and although the implementation of a water governance model from the perspective of Integrated Water Resources Management (IWRM) seems quite distant, there is the possibility of launching public actions from the territories from which networks are structured, articulate interests, and undertake collective actions [10].

3. The Dublin declaration

In the Dublin statement, according to international meetings on water resources management, it is recognized that water is essentially an economic good. Economists have insisted on the need to recognize that water is an economic good, so it should be possible to govern its distribution through the market [11]. There are a large number of approaches that explain the difficulty in improving water management systems, among them we have the lack of methods to design strategies that allow moving step by step from an existing situation to a desired situation [5].

The Dublin principles and the coherent action plans were collected at the Earth Summit held in Rio de Janeiro, in Chapter 18 of Agenda 21 [12], in the following terms: "A global ordination of fresh water as a limited and vulnerable resource and the integration of sectoral water plans and programs, within the national economic and social policy, are measures that are of the utmost importance among those adopted in the 1990s." There are controversies over the issue of the results of changes in water governance. This, we believe, can best be characterized as an effect identification problem and concerns the question of the effects of changes in governance on a wide variety of empirical phenomena and associated policy issues [13].

4. International instruments on groundwater use

It can be said that rarely, international law has taken into account groundwater, being known that there are many surface water treaties, groundwater is not nominally included in the scope of these instruments, especially if it is "related "with territorial surface waters. Some legal instruments contain specific groundwater conditions and do not address it exclusively. FAO and UNESCO provide a variety of mandatory and non-mandatory instruments of international law in this important field [14].

Likewise, there are directives on water resources and environmental treaties that contain provisions on groundwater, as is the case of the United Nations Convention to Combat Desertification in those countries that suffer severe drought and/or desertification, particularly in Africa. There is also the United Nations Convention on the law of the uses of international watercourses for purposes other than navigation. These statements are important but are generalized to resources of all kinds and in the second case to international watercourses, not being specific to the groundwater issue, being a strategic basis to address it.

There are also interstate agreements on groundwater or with groundwater provisions. United States, Idaho - Washington in the special form of the Inter-institutional Agreement on Coordinated Management Aquifer Groundwater Pullman-Moscow [14], it is agreed that management of groundwater resources in the aquifer will be in accordance with the management plan of groundwater adopted by the Pullman Moscow Water Resources Committee (PMWRC) to the extent that such plan can be implemented and administered according to the laws of each state.

5. National instruments on groundwater use: Peruvian case

In Peru, the granting of rights to use groundwater is regulated in Title IV of the Water Resources Law (article 110). In the aforementioned law, two additional provisions are required: the first states that in case of cessation of the use, either permanently or temporarily, the holders of the rights must adopt the necessary security measures in order to avoid causing damage to the aquifer and third parties. The second indicates that it is the obligation of groundwater users to install and maintain piezometers in the quantity and separation determined as provided by the corresponding authority, information which must be communicated to the "Autoridad Nacional del Agua (ANA)" [15].

According to Supreme Decree D.S. N ° 080-84-AG of September 6, 1984, the underground water of the pampas of La Yarada, and the D.S. N ° 020-87-AG of May 1, 1987 gives an extension for two more years. By means of Ministerial Resolution No. 0555-89-AG/DGAS of December 5, 1989, the execution of works intended to extract groundwater in the pampas of La Yarada is prohibited, later by RM No. 696-98-AG of December 16 of 1998 and based on the hydrogeological study of the Pampas de La Yarada, the closure of the increased exploitation of groundwater in the aquifer is declared. With a ruling by the Constitutional Court through a ruling issued in file No. 1290–2002-AC/TC, it ruled on the declaration of closure and the non-application of regulations for the regularization of water licenses in the area declared closed.

Regional Ordinance No. 009-2004-CR/GOB.REG.TACNA, declared of regional interest the intangibility and conservation of groundwater and rough land in the state of the pampas of La Yarada. In 2006, by means of Supreme Decree No. 065-2006-AG, the conservation and preservation of the Caplina valley water resource was declared of public necessity and national interest, extending the closure to the entire Caplina aquifer, ratifying it in the year 2009 through RJ N ° 327-2009-ANA of June 15, which declares the aquifer of the Caplina river valley closed, where the La Yarada aquifer is included, the prohibition of execution of groundwater exploitation works and the provision that the Local Water Administrators are responsible for the control and surveillance of the aquifers that are in their respective jurisdictional areas.

The R.J. N ° 201-2010-ANA, of March 22, 2010, ratifies the measures for the conservation and preservation of groundwater, based on the study "Numerical Modeling of the La Yarada Aquifer", which concludes that there is an overexploitation of the aquifer and recommends maintaining the closure of the exploitation of the aquifer system. Finally, in 2015, the D.S. N ° 007-2015-MINAGRI, in its second complementary provision establishes that "The areas declared closed prohibit their condition, proceeding exceptionally and for the only time to formalize or regularize water use licenses, in accordance with the provisions of this Supreme Decree".

6. The crisis of governance and governability

Governance refers to the management of collective affairs and involves the articulation of rules of conduct and the agreement of principles for the allocation of resources within the framework of a political community [16]. Likewise, governability has to do with the choice of goals and values that should guide society [17]. In Peru, we can affirm that there is no dispersion of legal provisions, nor a profusion of laws, which if we estimate is to find both conceptual and strategy and action gaps, which are really important. In most cases, the provisions are outdated and unrelated to the regional and national reality. Likewise, we currently have important instruments such as the integrating principle in water management, which is absent in the government actions taken to face the problem of the La Yarada aquifer.

The fragmentation of the actions regarding water management in the aquifer system is a relevant problem, the institutions act in isolation and without coordinating the joint actions necessary to treat the problem. This denotes that there is a governing body of water management policies in the country such as the ANA, which is perceived to have no technical and regulatory solvency to execute current water legislation on the La Yarada aquifer. The regional and local government has no participation in the management and territoriality of the water in the aquifer system.

There are limitations to the effectiveness of governance and it lies in the relationship of groundwater with society, there is a conception of water with inexhaustible resources, a culture of non-payment for water and the lack of responsibility for environmental protection. Regarding public policies and their incorporation strategies, these are practically nil, the fragmentation of policies is notorious and most importantly, there is fragility of state institutions, related to the issue of groundwater, for the control of use and its exploitation. The lack of government planning to achieve an adequate use of water is notorious. We can also identify a lack of education on the preservation and sustainable use of groundwater in the La Yarada aquifer system.

The management of the system is a latent problem, the legislation on penalties for water use crimes without authorization or without licenses or permits is not clear. There is no reliable inventory of existing wells with and without using authorization; therefore a water balance cannot be specified by region and considering levels of vulnerability. The incorporation of municipalities, governorates, and other institutions in the management of groundwater is sought through the formation of the Caplina - Locumba Water Resources Council, created by Supreme Decree No. 019-2013-AG, as a space for agreement in which the institutions and organizations of the region, linked to the integrated management of water resources can present their needs, projects and claims, in order to plan and coordinate the sustainable use of water in the basin. In the constitution of this council, serious conformation problems have been reported, some of its members, especially the president of the council, do not meet the requirements indicated in Law No. 29338, the Water Resources Law and its regulations, this situation ends up establishing a climate of mistrust regarding non-compliance with the law from the same governing entity of the use and conservation of water in the country, that is, the ANA.

7. Governance and governance models

Improving water governance is the key to address water insecurity in developing countries. The literature does not pay attention to the study of incentive structures, interdisciplinarity and orientation with clear political implications [18]. In the Fung and Wright model [19], the governance approach is presented as a non-hierarchical form of government, characterized by cooperation with non-state actors within mixed decision-making networks between the public and the private [19].

The UNESCO model of water governance is a theoretical model that focuses on government issues. The purpose of this is the formulation of policies and their execution by the State. The model is integrated by the dimensions, social, economic, of political training, and environmental sustainability [20]. In Musetta's hydraulic leviathan model [20], emphasis is placed on a centralized State, strong in terms of state presence, a State that plans and develops, that builds large works of hydraulic infrastructure (dams, dams, irrigation systems) and in this measure the potential of your strength. It is also a State that organizes the other actors in the economy and society, and never delegates the management of this position [21].

Guhl's hard and soft line model [22] are a theoretical model, in the hard line, focuses on increasing the supply of water and making it more stable in time and space, through the construction of infrastructure works that allow us to have a greater and more stable supply of water throughout the year, or availability in areas with deficits. Its fundamental objective is the conservation of the water supply and its efficient use; it gives comprehensive consideration to the hydrological cycle based on changes in attitudes and behaviors of the users regarding the valuation of water as a finite resource and the consequent limitation of its availability and development and use of more efficient technologies [22].

8. Proposal for the adoption of a governance model: La Yarada aquifer case

As mentioned, there are several models applied in different parts of the world, each with certain peculiarities. Based on the characteristics of the La Yarada aquifer system (**Figure 2**), due to the characteristics of the state's intervention in the economic activities of the region and the country, we can establish that the Mussetta hydraulic leviathan model is the best suited to the particularities of the aquifer in question. This is a strong model in terms of state presence, plans, developments, and works of hydraulic infrastructure. It is the State that in turn organizes the other actors in the economy and society, and never delegates the management of this position. This type of action is required in a system such as the one studied where there is a general disorder, regulations and laws are not complied with, the stability of the aquifer system having been put in a chaotic situation, while the recharge is much lower than the exploitation because unauthorized users have been generated to whom no sanction has been applied.

A clear example of the weakness of the State and its institutions is registered with the grant of the D.S. N ° 007-2015-MINAGRI, which establishes that "the areas declared closed prohibit their condition, proceeding exceptionally and for the only time to formalize or regularize water use licenses, in accordance with the provisions of this Supreme Decree." This decree is totally contradictory, existing the RM N ° 696-98-AG that ratifies the closure in the La Yarada aquifer, the formalization of licenses for unauthorized wells is decreed, which has generated a great imbalance in the system, which has produced not only a decrease in the water table but also deterioration of water quality due to marine intrusion processes in a large part of the aquifer, as demonstrated by studies carried out by ANA.

The post emission experience of the D.S. N ° 007-2015-MINAGRI has shown that the situation worsened, in the period of time granted for the presentation of files, a massive and uncontrolled action was made to drill wells in the study area in

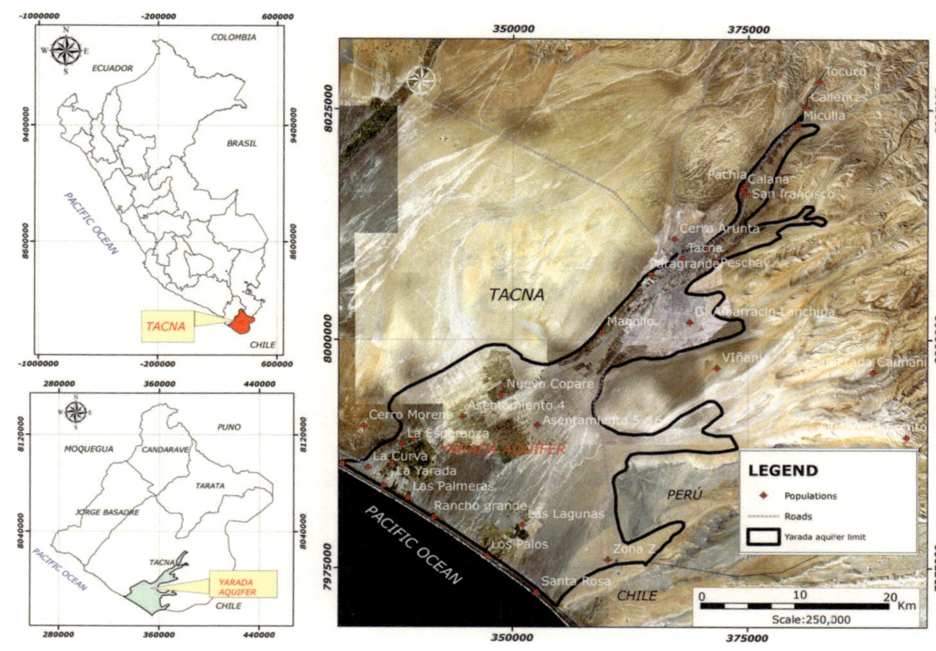

Figure 2.
La Yarada aquifer location map.

order to achieve the regularization of the groundwater exploitation infrastructure. Likewise, after the regularization process carried out, well drilling continues indiscriminately, with which it is demonstrated that the effect, more than a palliative to the problem, was a trigger for the drilling of wells without control, ignoring the regulations current and most worrying, an attempt was made to implement an action without the required control and security mechanisms.

According to the particularity of the La Yarada aquifer system and water management in Peru, it is observed, from the perspective of water governance, the presence and actions of social groups have increased over the years, exerting strong pressure on the aquifer. Thus, we can establish that the transition from governance to governance presents indicators that the institutions have not adequately managed their tasks and responsibilities and that the social agents have gained a presence in the excessive extraction of water from the aquifer, without the authority being able to intervene in concrete form. No sanctions are applied for violators of the laws and restrictive devices for the use of aquifer water, generating an increase in extraction wells that have grown geometrically in recent years with the consequent deterioration of the quality of the water in the aquifer.

9. Conclusions

Government policy in many countries of the world generates a serious collision between closed regulations on coastal aquifers and, on the other hand, decrees authorizing the formalization or regularization of water use licenses. This situation generates negative antecedents of government policy in the administration and management of water and in general of all-natural resources.

Governance and governance are altered and exceeded, the laws regarding the regimes of management and limitation of groundwater resources, both have qualities that could provide a relevant benefit for the sustainable use of water resources, if no immediate action is taken On the part of the government apparatus, the crisis of the system could become irreversible with the consequent economic damage that this entails. In the area, there are large extensions of plantations that are the support of economic activity.

In the Peruvian case, it is necessary to have a specific regulatory framework for groundwater that recognizes its particularity. Institutions in the field of Water Law must be consolidated and strengthened, not being a unique need in the case of groundwater, but in general for water resources in all its forms. Therefore, the efficient and transparent participation of the public administration that allows adequate water management for sustainable use is urgent. In the Peruvian case, it would be advisable to adopt the Mussetta hydraulic leviathan model, strong in terms of state presence, planning, developing, and building hydraulic infrastructure works. In turn, the State is the one that organizes the other actors of the economy and society, and does not delegate functions. This type of action is required in a system such as the one studied, where there is a general disorder, regulations, and laws are not complied with, the stability of the aquifer system being put in a chaotic situation, while the recharge is much less than exploitation because unauthorized users have been generated to whom no sanction has been applied.

Acknowledgements

This work has been developed within the framework of the research project "Integration of hydrodynamic, hydrochemical and isotopic methods to specify the operation and sustainable management of the Yarada aquifer, Tacna, Peru," financed with funds from canon, supercanon and mining reserves of the Jorge Basadre Grohmann National University, Tacna, Peru.

Author details

Edwin Pino[1]* and André Steenken[2]

1 Civil Engineering Department, Faculty of Civil Engineering, Architecture and Geotechnics, National University Jorge Basadre Grohmann, Tacna, Perú

2 Geological-Geotechnical Engineering Department, Faculty of Civil Engineering, Architecture and Geotechnics, National University Jorge Basadre Grohmann, Tacna, Perú

*Address all correspondence to: epinov@unjbg.edu.pe

References

[1] Mohamed E, Lashin A, Abdalla F, Al-Bassam A. Assessing the hydrogeochemical processes affecting groundwater pollution in arid areas using an integration of geochemical equilibrium and multivariate statistical techniues. Environmental Pollution. 2017;**229**:760-770

[2] Todd F. New NASA data show how the world is running out of water. The Washington Post. USA. 2015. Available from: https://www.independent.co.uk/environment/nasa-data-shows-the-world-is-running-out-of-water-10325188.html [Accessed: 10 April 2018]

[3] Minaverry C. Consumidores y usuarios del servicio del agua en Argentina. Enfoques jurídicos. Tecnologia y Ciencias del Agua. 2017;**8**(1):5-20

[4] Villalobos O. Cuál es la diferencia entre gobernabilidad y gobernanza Journal Rural.com. 2015. Available from: http://journalrural.com/gobernabilidad-y-gobernanza-local/#more-2407 [Accessed: 15 April 2018]

[5] Dourojeanni A, Jouravlev A. Crisis de Gobernabilidad en la Gestión del Agua (Desafíos que enfrenta la implementación de recomendaciones contenidas en el cap. 18 del Programa 21). CEPAL. Santiago de Chile. Chile. 2001. p. 83

[6] Rodríguez M. La posible creación de mercados de agua y la gobernabilidad de este recurso en Colombia. Comentarios sobre el proyecto de ley del agua. Revista de Ingeniería. 2005;**22**:94-102

[7] Retamal M, Andreoli A, Arumi J, Rojas J, Parra O. Gobernanza del Agua y Cambio Climático: Fortalezas y Debilidades del Actual Sistema de Gestión del Agua en Chile. Análisis Interno. Interciencia. 2013;**38**(1):8-16

[8] Zamudio C. Governance of water resources in Colombia: Between progress and challenges. Gestión y Ambiente. 2012;**15**(3):99-112

[9] UNESCO. Gobernanza de las aguas subterráneas en los acuíferos transfronterizos (Proyecto GGRETA). PANORAMA Y RESULTADOS DE LA ETAPA DE EVALUACIÓN (2013-2015). París. 2015. Available from: http://unesdoc.unesco.org/images/0024/002430/243003s.pdf [Accessed: 15 April 2018]

[10] González N. Desafíos de la gobernanza ambiental: Una aproximación a las implicaciones de la Gestión Integrada del Recurso Hídrico en Colombia. Ciencia Política. 2017;**12**(23):205-229

[11] Terence R, Jouravlev A. Princes, property and markets in water allocaton, (LC/L 1097). Comisión para América Latina y el Caribe (Cepal). Santiago de Chile. 1998. Available from: https://repositorio.cepal.org/handle/11362/5735 [Accessed: 12 May 2018]

[12] WWAP. Water a shared responsibility. The United Nations World Water Development Report 2. UNESCO. France. 2006. Available from: http://unesdoc.unesco.org/images/0014/001444/144409E.pdf [Accessed: 15 April 2018]

[13] Kersbergen K, Waarden F. 'Governance' as a bridge between disciplines: Cross-disciplinary inspiration regarding shifts in governance and problems of governability, accountability and legitimacy. European Journal of Political Research. 2004;**43**:143-171

[14] Burchi S, Mechlem K. Agua subterránea en el Derecho internacional; Recopilación de tratados y otros instrumentos legales. FAO Estudio

legislativo 86. Roma. Italia. 2007. Available from: http://www.fao.org/docrep/008/y5739e/y5739e00.htm#Contents [Accessed: 20 April 2018]

[15] Cairampoma A, Villegas P. Régimen jurídico de las aguas subterráneas en el Perú. THĒMIS-Revista de Derecho. 2016;**1**(69):147-158

[16] Healey P. Collaborative Planning: Shaping Places in Fragmented Societies. Canada: UBS Press. Vancouver; 1997. p. 339

[17] Green S, Peterson W. Governance in Contemporary, the Semisovereign State Revisited. Vol. 334. New York. USA: Cambridge University Press; 2005

[18] Silva J. Propuesta de un modelo de gobernanza del agua. XX Congreso Internacional de Contaduría, Administración e Informática de la UNAM. México. 2016. Available from: https://www.researchgate.net/publication/287207692_Propuesta_de_un_modelo_de_gobernanza_del_agua [Available from: 12 May 2018]

[19] Fung A, Wright E. Deepening Democracy: Innovations in Empowered Participatory Governance. London-New York: Verso; 2003. p. 314

[20] UNESCO. El agua, una responsabilidad compartida. Segundo informe de las Naciones Unidas sobre el desarrollo de los recursos hídricos en el mundo. 2006. Available from: http://unesdoc.unesco.org/images/0014/001444/144409S.pdf [Available from: 01 April 2018]

[21] Mussetta P. Participación y gobernanza. El modelo de gobierno del agua en México. Espacios Públicos. 2009;**12**(25):66-84

[22] Guhl EH. Una Gestión Integrada del Agua en la Región Andina. Bogotá. Colombia: AECID; 2008. p. 168

Chapter 3

Groundwater Recharges Technology for Water Resource Management: A Case Study

Jatoth Veeranna and Pawan Jeet

Abstract

The irregularity in monsoon has severely affected the water availability at surface and sub-surface systems. Diminishing surface and sub-surface availability has not only decreased the water availability, but it additionally affected the ecosystem and increased disastrous situations like floods and droughts, resulting problems of stress on groundwater recharge. Groundwater recharge is a technique by which infiltrated water passes through the unsaturated region of groundwater and joins the water table. It is based upon soil type, land use land cover, geomorphology, geophysical and climate (viz. rainfall, temperature, humidity etc.) characteristics of a region. Over the years, due to variations in weather pattern and overexploitation of aquifers groundwater recharge has decreased and groundwater level has reduced in the most parts of the country. This has led to severe water deficit problems in several parts of the country. This can be solved by different direct and indirect methods of groundwater recharge technology. This technology can reduce the wastage of water and enhance groundwater availability for uses in different sector like irrigation, domestic and industrial uses.

Keywords: groundwater recharge technology, groundwater, rainfall, water harvesting

1. Introduction

Groundwater is the foremost supply of freshwater that caters to the demand of household, agricultural and industrial sectors. It has become an essential for domestic uses especially for drinking water and food security for billions of population of the country. Approximately 70 percent water use in rural regions and approximately 50 percentage of water used in urban and commercial region are fulfilled by the sources of groundwater [1]. The average annual per capita water availability has been steadily falling since 1991 (2300 m^3) to 2015 (1720 m^3) in the country and these are projected to reduce to 1400 m^3 and 1190 m^3 for the years 2025 and 2050, respectively [2]. In last three decades an exponential growth in number of ground water structures has been observed, leads to enormous withdrawal of groundwater for various uses in different sectors. However, speedy urbanization and land use adjustments has resulted in decreased natural infiltration/recharge of aquifers. This has result in various issues related to quantity and quality of groundwater, the decline in water table levels and depletion of groundwater resources [3].

Slow natural replenishment of groundwater reservoir is not able to keep pace with the excessive persisted exploitation of groundwater assets in numerous parts of the country. In order to increase the natural supply of groundwater, artificial recharge to groundwater has emerged as a vital and frontal management approach [4]. So, the recharge of groundwater is performed by using various direct and indirect recharge technologies. The adoption of this technology depends on hydrology, geology and other factors of a region [5]. Thus artificial recharging of aquifer is one of the best options, in order to improve groundwater crises which are sustainable in the long term.

2. Need to recharge groundwater

Artificial recharge is the process by which the ground water recharge is augmented at the rate much higher than those under natural condition of percolation.

According to [1] groundwater recharge meets the demand as

- By 2050 the demand for ground would reach to 1180 billion cubic meters.
- Water will be available to 1 person out of 3 and no food to 1/3rd population

In the country, either in pre- monsoon or post-monsoon or both the annual groundwater yields exceeds the annual groundwater recharge, resulting a significant fall in long term groundwater levels has been observed.

3. Problems associated with groundwater overexploitation

3.1 Lowering of the ground water level

The most critical effect of over-pumping of groundwater is that the water table may be lowered. For water to be withdrawn from the aquifers, water should be pumped from a well that reaches beneath the water table. If groundwater levels reduce too far, then the well owner may have to deepen the well, drill a new well, or, at least, lower the pump below the water table available. Also, as water levels fall down, the rate of water the well can yield might also reduce.

3.2 Increased costs for the user

As the depth of groundwater increases, the water must be lifted from the depth of groundwater to the ground surface. If water pumps are used to lift the water, more electrical energy is required to operate the water pump. Under this situation, using the well can become costly to users.

3.3 Reduction of water availability in water bodies

There may be more of an interaction between the water in water bodies such as ponds, lakes, rivers and streams, and groundwater than the most of the people think. A proportion of water flowing in rivers contributed from seepage from the streambed to groundwater. Groundwater contribution mainly depends upon on the parameters of physiographic, region's topography, soil, geology and climate.

3.4 Land subsidence

The primarily cause of land subsidence is a lack of support underneath ground surface. Sometimes, when groundwater is over-exploited, the soil collapses, compacts, and sinks. This depends on various factors, such as the type of soil, soil compressibility, physical attributes of the aquifer, water table levels and earth geology. It is most often caused by anthropogenic activities, mainly from the excess removal of subsurface water.

3.5 Deterioration of water quality

Saltwater intrusion is the major cause and threat to contamination of fresh groundwater supplies. Available volume of water in the aquifers is not fresh water; much of the very deep groundwater and water below seas is saline. Under natural conditions the boundary between the freshwater and saltwater tends to be relatively stable while under excessive pumping conditions it may result saltwater to migrate inland and rising upward and it leads to contamination of the water supply.

4. Benefits of groundwater recharge

There are following advantages of artificial recharging of groundwater aquifers:

- Subsurface storage space is available free of cost and inundation is avoided
- Evaporation losses are negligible and temperature variations are minimum
- Quality improvement by infiltration through the permeable media
- It has no adverse social impacts such as displacement of population, loss of scarce agricultural land etc.
- It is a environment friendly technology that controls soil erosion and flood like situations, and provides sufficient soil moisture during dry spell or water deficit conditions.
- Water stored in soil profile is relatively immune to natural and man-made catastrophes.

5. Artificial groundwater recharge technology

The artificial groundwater recharge technology can be broadly categorized as follows [6].

Direct technology			Indirect Technology
Surface	**Sub-surface**	**Combination surface and sub-surface**	
• Flooding/ Water spreading • Basins or percolation tanks • Stream augmentation • Ditch and furrow system • Over irrigation	• Injection wells or recharges wells • Recharge pits and shafts • Dug well recharge • Bore hole flooding • Natural openings, cavity fillings.	• Basin or percolation tanks with pit shaft or wells.	• Induced recharge from surface water source • Aquifer modification

Except above, water conservation structures like dams, sub-surface dykes (or locally termed as Bandharas) are entirely prevalent to capture sub-surface flows. Similarly, in hard surface areas rock fracturing strategies such as sectional blasting of boreholes with suitable techniques has been operated to inter-connect the fractures and gear up the groundwater recharge. Cement sealing of fractures, through specially built borewell has been utilized in the state of Maharashtra to preserve sub-surface flow and increase borewell yield [7].

6. Direct surface techniques

This method of groundwater recharge is very simple and most widely used. Under this method stored surface water is directly conveys into an aquifer without infiltration and water percolates naturally through the unsaturated zones of soil profile and join the groundwater table.

6.1 Flooding/water spreading

This is a very common method of groundwater recharge (**Figure 1**) [8]. This is

- This method is suitable for relatively flat topography
- Water is spread as a thin sheet
- Higher rate of vertical infiltration is obtained
- Potential area for this method is alluvial region of country

6.2 Percolation tank/basin

- A percolation tank can be defined as an artificially created surface water body in a highly permeable land submerged area so that the surface runoff is made to percolate and recharge the groundwater storage (**Figure 2**) [9].
- It is the most prevalent structures in India because it is used to measure the recharge the groundwater reservoir in highly permeable land areas.

Figure 1.
Water spreading technique.

Figure 2.
Percolation tank for water storage.

- It is applicable in both alluvial as well as hard rock formations regions.
- Its efficacy and feasibility is more in hard rock formation regions than alluvial regions.
- Suitable in the States of Himachal Pradesh, Jammu & Kashmir, Uttrakhand, Sikkim, North Eastern States, Punjab, Haryana, Uttar Pradesh, Bihar, West Bengal, Madhya Pradesh, Maharashtra, Karnataka, Tamil Nadu, Andhra Pradesh, Kerala

6.3 Stream augmentation (Check dams/Nala bund/ gabions)

- It is feasible to construct across small streams having gentle slope (less than 6 percent) [6].
- It is applicable in both hard rock as well as alluvial formation region.
- It is mainly confined to stream course and its height is normally very less (less than 2 m).
- To harness the maximum run off in the stream, series of such check dams can be constructed (**Figure 3**).
- A nala bund acts like a mini percolation tank.
- These are popular and feasible in Bhabar, Kandi and talus scree areas of Uttar Pradesh, Punjab, and Maharashtra.

Figure 3.
Check dam for surface water harvesting.

6.4 Ditch and furrow technology

This technology is mainly suitable in areas of irregular topography, shallow and flat bottomed and closely spaced furrows or ditches that provide more surface area under groundwater recharge through canal, river, stream and so on. This requires fewer earthworks and also less sensitive to siltation.

7. Direct sub-surface techniques

7.1 Dug well recharge

- It is suitable in alluvial as well as hard rock areas having depth upto 50 meters (**Figure 4**) [10].
- The ground water reservoir, storm water, tank water, canal water etc. can be diverted into these structures to directly recharge the dried aquifer.
- Ordinary dug wells, borewell and tube wells can be used for recharging of gw recharge takes place by gravity flow.
- Suitable in Madhya Pradesh, Maharashtra, Karnataka, Tamil Nadu, Andhra Pradesh, Kerala.

7.2 Recharge pit and shaft

- These are the most efficient and cost effective structures to recharge the aquifer directly (**Figure 5**) [6].
- In area where impervious layer is encountered at shallow depth.
- Where phreatic aquifer is not hydraulically in connection with surface water.
- The diameter of shaft should be more than 2 m for recharging rate 7–14 lps.

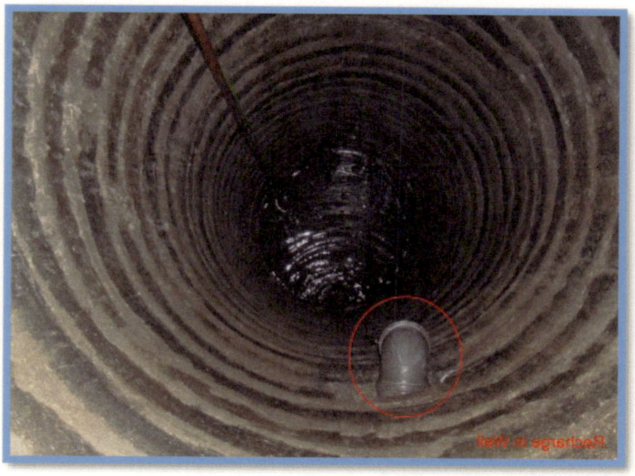

Figure 4.
Dug well groundwater recharge structure.

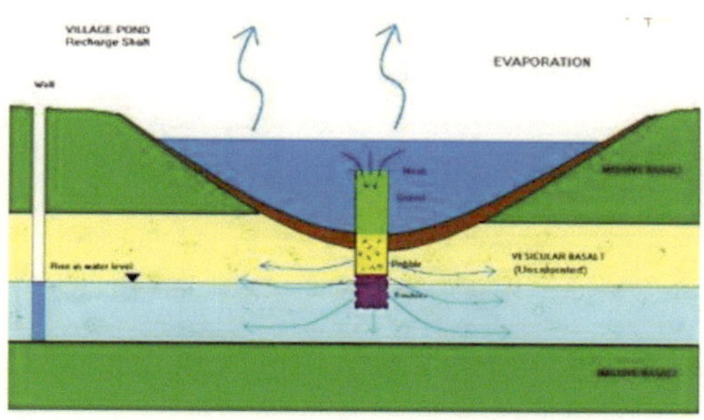

Figure 5.
Recharge pit for groundwater recharge.

- These structures are common in the states of Maharashtra, Madhya Pradesh, Andhra
- Pradesh, Bihar, Gujarat, Himachal Pradesh, Jammu& Kashmir.

7.3 Injection wells

Injection well is generally recommended in urban areas (**Figure 6**). It is operational in certain hydrogeological setting for groundwater recharge where the aquifers do not get the natural recharge because of the confining layers of low permeability [11]. There are following advantages for installation of the injection well:

- It is made with the purpose of augmenting the ground water storage of a confined aquifer by pumping-in treated surface water under pressure.
- The aquifer to be replenished is generally over-exploited.
- It is suitable in coastal regions to capture sea water and also to withstand the land subsidence problems in the regions where confined aquifers are over-pumped.

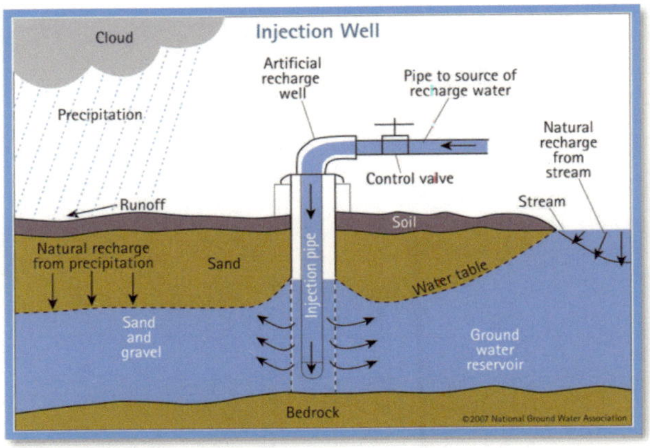

Figure 6.
Injection well for groundwater recharge.

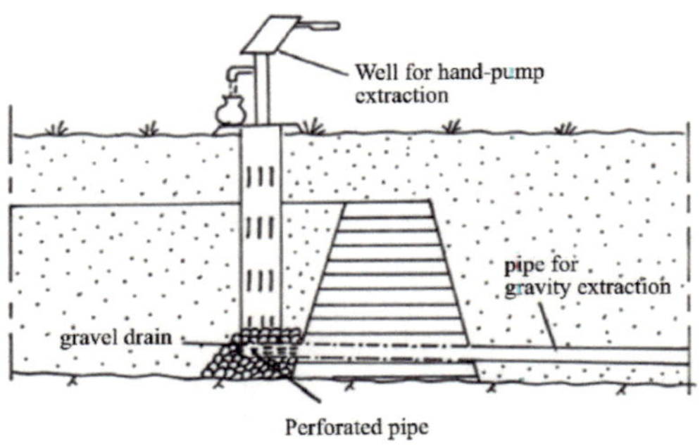

Figure 7.
Subsurface dyke for surface water harvesting.

- Water available for groundwater recharging is to be fairly treated for elimination of suspended material, chemical stabilization and bacterial manipulation.

7.4 Subsurface dykes

- It is a sub-surface barrier across a stream which slows down the natural sub-surface /groundwater flow of the system and capture water beneath ground surface to meet the water demand (**Figure** 7) [12].
- The main cause of groundwater dam is to capture the flow of groundwater out of the sub-basin and increase the storage capacity of the aquifer.
- Suitable in hard rocks or alluvium forested area.

8. Case study

Direct and indirect method of groundwater recharge technology was adopted in different locations of India.

Artificial groundwater recharge system at Shram Shakti Bhawan, New Delhi.

Shram Shakti Bhavan having catchment area of about 11,965 km^2. Roof top rainwater harvesting may annually recharge about 2900 m^3 water this is going as waste. Over the years, depth to water level has fallen by 2 to 3 m in the catchment area. Artificial groundwater recharge is proposed via recharge trenches with two injection wells in each at selected locations in catchment area [10].

8.1 Salient features

- Rainwater runoff: 3325 cubic meter
- Area of 11,965 sq.km
- Recharge Structures Trench & recharge wells: 3
- Year of construction: 2001
- Average recharge: 3000 m^3/Year
- Rise in water levels Aug 2007: 1.68–3.33 m
- Cost: Rs 4.10 Lakh

9. Artificial groundwater recharge system at President's Estate, New Delhi

President's Estate having catchment area of about 1.20 km^2 is located on the Northern flank of Delhi ridge. Extravagant groundwater development has resulted in annually groundwater fallen by 6 to 13 m. Four metre thick aquifer has become de-saturated over an area of 0.7 km^2. The artificial groundwater recharge in the catchment area is being done through two dried dug wells, one vertical recharge shaft, one injection well, two recharge trenches with injection wells and so on [10].

Annually approximately 28,170 m^3 rainfall-runoff water collected and used to recharge groundwater (**Figure 8**).

Figure 8.
Artificial groundwater recharge structure at President's estate.

9.1 Salient features

- Campus area: 1.3 Sq.km
- Source of water: Rainwater & Swimming pool water
- Av annual rainfall:712.2 mm
- Depth of water table: 6–12 m. Bgl
- Year of construction: 2000
- Recharge structures:
    - Two existing dug wells
    - One recharge well
    - One recharge shaft
    - Two trench with recharge well

Rise in water level during 2003: Upto 4 m.

9.2 Artificial Recharge Structures (Sub surface Dykes) in Rajgarh district, M.P.

- District: Raigarh, M. P.
- Normal Rainfall: 950 mm
- Location: Barwa Kalan, Ajnar subbasin, Rajgarh district
- Type of Structure: Subsurface Dykes
- Cost: Rs. 2.0 lakhs
- Implementing Agency: Public Health Engineering Department, Govt. of M.P.
- Geology: Alluvium and basalt
- Year of completion: 2004–2005
- Study year: 2009–2010

Description: To capture the over groundwater flow to the river Ajnar, in the form of base flow, subsurface dykes were constructed near Barwa Kalan village (**Figure 9**) [10].

9.3 Impact

- Water level in the nearby dug wells recorded a rise in the range of 0.80 to 3.80 m.
- In the hand pumps, the water level rise in the range of 6.0 to 12.0 m.

Figure 9.
Artificial groundwater recharge structure at Rajgarh district in MP.

- During the period the cultivated areas under *rabi* crop has increased from 97 to 121 ha.
- The number of irrigation wells increased up to 102 from initially 38.

10. Conclusions

Artificial groundwater recharge technology is very impressive technology to increase water table and groundwater availability. It plays an important role in the reduction of surface runoff, increase availability of water for irrigation, domestic and industrial sector, improve the drainage, revival of springs and improvement of groundwater quality and so on. It is also considered to mitigate the impacts of variability in rainfall patterns under varying climatic conditions. Additionally, it is primarily important to meet the demand of spatial water productivity and availability at regional and global scale.

Author details

Jatoth Veeranna[1] and Pawan Jeet[2*]

1 Prof. Jayashankar Telanagana State Agricultural University (PJTSAU), Hyderabad, Telangana, India

2 ICAR Research Complex for Eastern Region, Patna, Bihar, India

*Address all correspondence to: pawan.btag@gmail.com

References

[1] Central Water Comission. Water and related statistics. 2007; Central Water Commission, Ministry of Water Resources, Govt of India, New Delhi.

[2] Press Trust of India. India's per capita water availability to decline further: ICAR. The Hindu News, 2019; https://www.thehindubusinessline.com/economy/agri-business/indias-per-capita-water-availability-to-decline-further-icar/article29342714.ece#

[3] Ministry of Rural Development. Water Harvesting and artificial Recharge. 2004.

[4] Kambale JB, Sarangi A, Singh DK, Singh AK. Performance Evaluation of filtration unit of groundwater shaft: groundwater study. *Current Science*, 2009; 96(4): 471-474. https://www.jstor.org/stable/24105453.

[5] Alderwish AM. Induced recharge at new dam sites- Sana'a Basin, Yemen. Arab *J. of geosci*, 2010; 3: 283-293.

[6] CGWB. Guide on artificial recharge to Groundwater: 2007; 1-54 (http://www.cgwb.gov.in/documents/Manual-Artificial-Recharge.pdf)

[7] Palanisami K, Raviraj A, and Thirumurthi S. Artificial Recharge in Hard Rock Areas of Coimbatore District –A Case Study. International conference on Groundwater sustainable Development problems, perspectives and challenges (IGC-2006). 2006. http://jnuenvis.nic.in/publication/IGC%202006%20Abstracts.pdf.

[8] Silber-Coats N, and Eden S. Water Banking, Recharge, and Recovery in Arizona, Arroyo. University of Arizona, Water Resources Research Center, Tucson, AZ, 2017; 2-16.

[9] Chaitanya. Percolation Tanks – A Component for Ground Water Recharging. Chaitanya Rain Harvest Products & Systems Pvt. Ltd., 2018. http://www.chaitanyaproducts.com/blog/percolation-tanks-a-component-for-ground-water-recharging/.

[10] CGWB. Case studies on rainwater harvesting and artificial groundwater recharge: 2011; 1-200.

[11] Hernandez J. Monitoring water disposal wells use WirelessHART transmitters: 2014; Industrial Ethernet Book Issue 92 (9).

[12] Norouzi S. Suggested Strategies in Water Treatment by Using Situ Pressure in Reverse Osmosis. *Open Journal of Geology*, 2015, 5: 367-373. **DOI:** 10.4236/ojg.2015.55033.

Chapter 4

Cumulative Groundwater Impact Assessment and Management – An Example in Practice

Sanjeev Pandey, Randall Cox and Steven Flook

Abstract

Production of coal seam gas (CSG), or coal bed methane, requires large-scale depressurisation of a target formation by extracting groundwater, which, in turn, has the potential to affect overlying and underlying aquifers. This leads to wide-ranging stakeholder concerns around the impacts on groundwater assets such as water supply bores, groundwater-dependent ecosystems and connected water-courses. Around 2010, the CSG industry in Queensland, Australia grew rapidly with the expansion of operations in the Surat and Bowen basins by multiple operators. This particularly raised concerns about the cumulative effects, because the target coal seams are part of the Great Artesian Basin – one of the world's largest aquifers. To respond to this challenge, an innovative framework was developed to provide for an independent cumulative impact assessment and to set up arrangements for managing those impacts. This chapter describes the main thrust of that framework.

Keywords: coal seam gas, cumulative impact assessment, groundwater management, Great Artesian Basin, groundwater, Queensland, regulatory framework, Surat Basin

1. Introduction

In the Surat Basin of Queensland, Australia, production of coal seam gas (CSG), or coal bed methane as it is known in the Americas, requires extraction of groundwater to depressurise the Walloon Coal Measures (the target formation). CSG has grown to become the dominant source of natural gas in Queensland, Australia, comprising more than 95% of the gas produced and more than 99% of the remaining proved and probable gas reserves [1]. The CSG produced from the Surat and Bowen basins is the feed stock for the liquefied natural gas export industry based in Gladstone. By 2009 to 2011, an unprecedented scale of CSG development was proposed in environmental impact statements (EIS) by four major proponents – Santos Limited, Australia Pacific LNG Pty Limited., QGC Pty Limited and Arrow Energy Pty Ltd – whereby a maximum of about 34,000 CSG wells were proposed in an area of about 37,000 km^2 [2].

The target formation for CSG production in the Surat Basin is part of the Great Artesian basin (GAB) – one of the largest groundwater systems in the world. This geology raised issues surrounding impacts of this development on groundwater assets such as water supply bores, groundwater-dependent ecosystems (GDE) and connected watercourses. There are an estimated 22,000 water supply bores in and around the CSG development area, along with a number of ecologically significant springs.

Exploration for CSG commenced in Queensland in the 1980s. Commercial production from the late Permian coal seams of the Bowen Basin commenced around 1995. By the early 2000s, the focus for development had shifted to the overlying Surat Basin, a part of the GAB (**Figure 1**), targeting the Jurassic age Walloon Coal Measures. However, over time, the development plans were revised downward in response to emerging market conditions and resource availability. Based on current

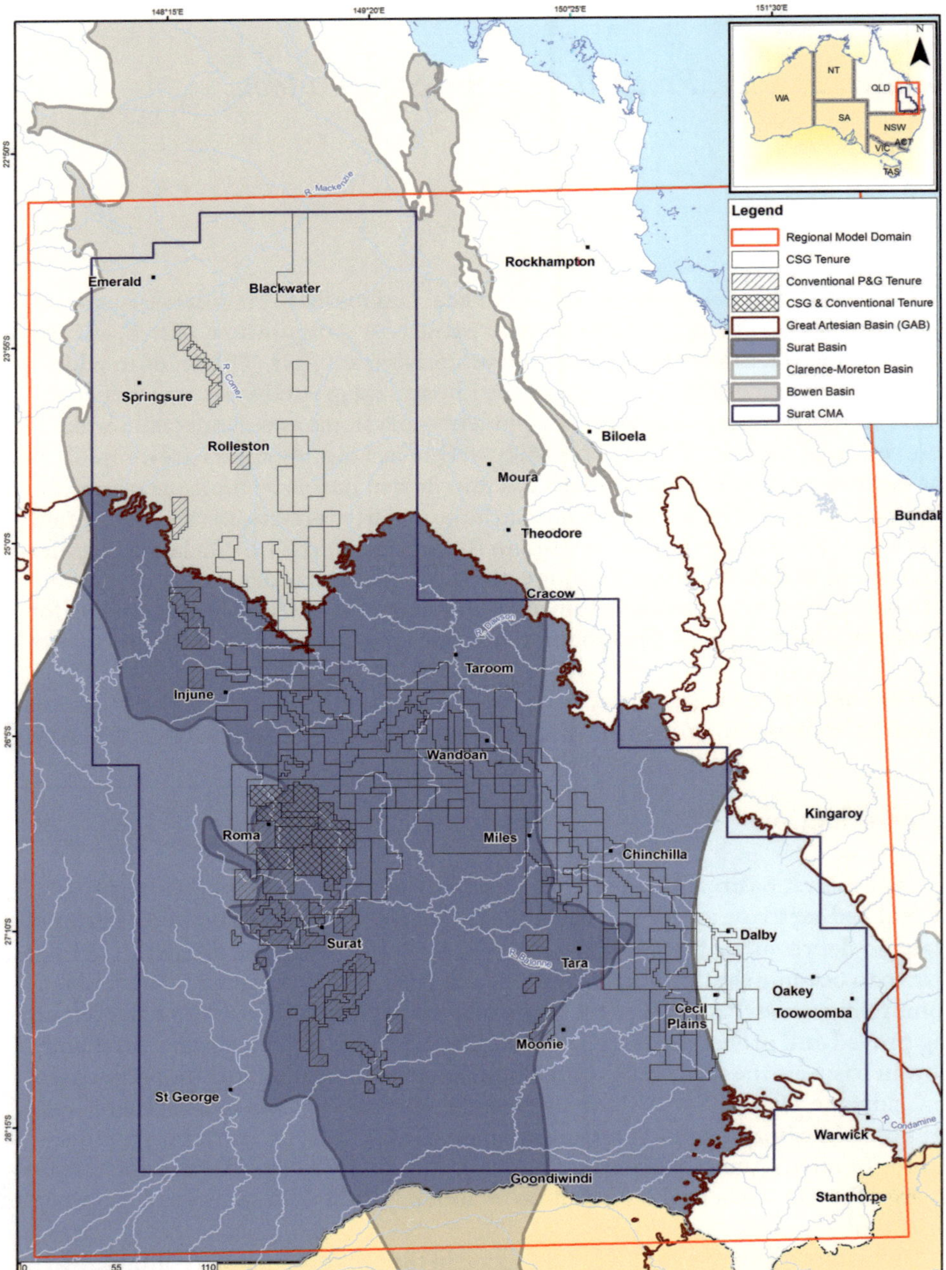

Figure 1.
A map showing the Surat Cumulative Management Area boundary in relation to the Surat Basin and the CSG tenures.

development plans, an estimated 21,000 CSG wells will have been constructed by the end of the life of the industry, of which about 7,000 are already in operation [3], extracting some 60,000 ML/year of groundwater from the CSG target formations in the Surat and southern Bowen basins.

This chapter provides some contextual background on the concept of cumulative impacts and related issues, followed by a description of how a framework was developed and implemented for managing those issues. The framework is a good example of proactive and adaptive groundwater management covering a cycle – from identifying issues to assessment and modelling, reporting, implementation and monitoring.

2. Context

2.1 The concept of cumulative impacts

The concept of cumulative impacts in water and environmental management is not new, but the context in which it is used varies widely. The term is sometimes used in relation to impacts of a single project on multiple social, environmental and economic factors [4]. In other instances, it is used to refer to impacts from the interaction of multiple activities, and/or the collective impact of many similar activities over time and space [5]. In this chapter, the term 'cumulative impacts' refers to groundwater pressure impacts from multiple CSG projects.

Regional or strategic assessments are often seen as mechanisms to assess cumulative impacts. Many authors have argued that cumulative effects are best assessed in a more regional and strategic context, at the level of strategic environmental assessment (for example, [6–9]). The Government of Alberta, Canada has developed a regulatory framework to better manage cumulative environmental effects from development through a regional planning instrument [10].

Although the term 'cumulative impacts' is not always used explicitly, the combined effects of all consumptive water use have always been considered, typically in catchment-scale and/or aquifer-scale planning for the allocation and management of water resources in Australia, following the 1994 Council of Australian Governments water reform agenda.

2.2 The Surat Basin

The Surat Basin underlies 180,000 km^2 of southeast Queensland. It is connected to the Eromanga Basin to the west, the Clarence-Moreton Basin to the east and the Mulgildie Basin to the northeast. It is a Jurassic to Cretaceous age sequence of alternating sandstones, siltstones and mudstones, with coal seams of economic significance in some areas (**Figure 2**, Video 2) [3]. The Surat Basin overlies the Permo-Triassic Bowen Basin sediments and is overlain by inliers of Quaternary alluvium and Tertiary basalts, particularly in the east. The total thickness of the Surat Basin sequence is about 3,000 m, with sediments deposited on an older erosional surface of the Bowen Basin.

The outcrop is the recharge area, with groundwater flowing generally along the formation dip for the deeper aquifers in confined areas, although there is a significant proportion of flow in outcrop areas northward along the topographic elevations [11, 12]. For the most part, groundwater in the Surat Basin occurs under sub-artesian conditions. Artesian aquifer conditions are only encountered in the southwest corner.

There are some 22,400 private water supply bores within the GAB footprint of the Surat Basin, the equivalent Clarence-Moreton Basin and adjoining parts of the southern Bowen Basin [3]. A majority of these bores access groundwater from the shallower unconsolidated alluvium or tertiary formations. The underlying GAB formations are primarily accessed for agriculture, town water supply and stock and domestic use, totalling 41,000 ML/year.

GDEs, which are associated with springs and baseflow-fed streams and include deep-rooted terrestrial vegetation, occur within the area. Springs are known to source water from some of the GAB formations. Natural groundwater discharge along the outcrop areas also feeds into watercourses and supports flow in dry periods.

2.3 Groundwater management challenges

Groundwater in the Surat Basin has long been accessed primarily for consumptive use. Long-standing arrangements for managing groundwater had been designed for consumptive use. The rapid emergence of the CSG industry, which extracts large amount of incidental groundwater (non-consumptive use) to depressurise coal seams, challenged those arrangements. For context, average CSG-related extraction over the life of the industry is expected to be about 51,000 ML/year [3], although these estimates have been declining over the years.

There are significant regional aquifers above and below the Walloon Coal Measures (**Figure 2**, Video 2) and therefore depressurisation for CSG production could potentially impact water supply bores and GDEs that rely on those aquifers (Video 3).

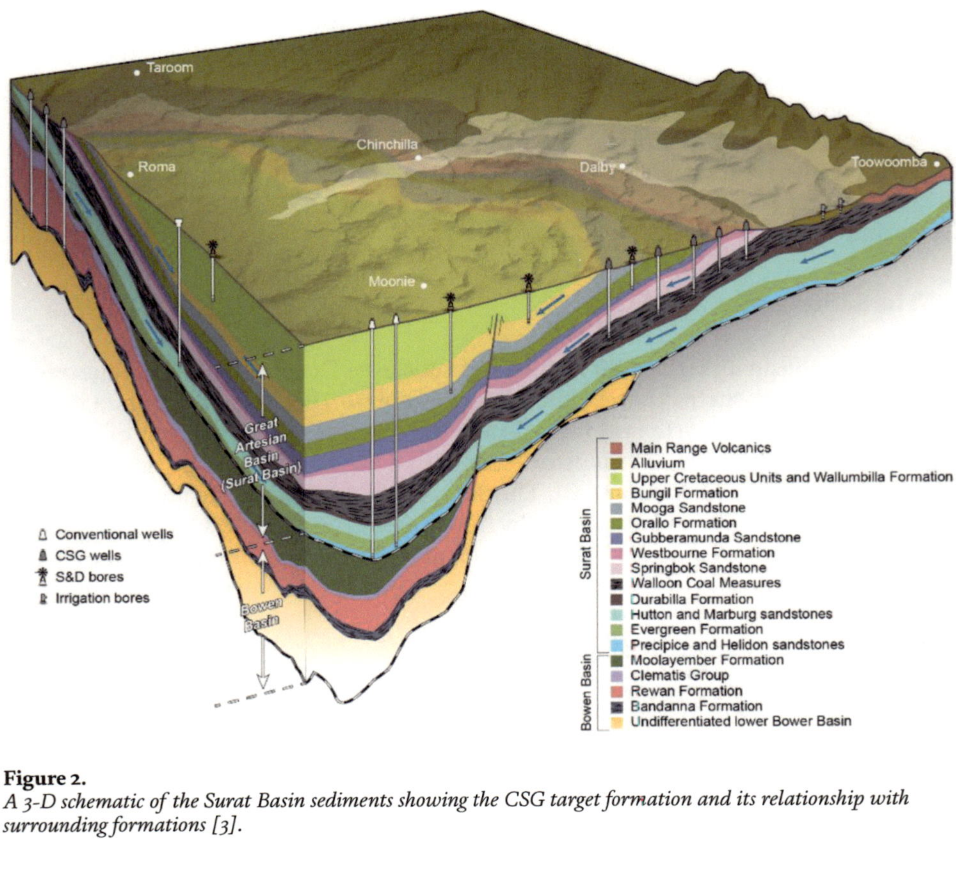

Figure 2.
A 3-D schematic of the Surat Basin sediments showing the CSG target formation and its relationship with surrounding formations [3].

Although the primary groundwater management concern was the management of the impacts of CSG water extraction on groundwater supplies and GDEs, there were also other concerns about non-pressure-related impacts on groundwater resources, such as: the potential for groundwater pollution from drilling activity; the beneficial use of the formation water extracted during development; and social impacts associated with the large workforce operating in the area [13]. These other concerns are beyond the scope of this chapter.

There was significant resistance to proposed CSG development from landholders and the community in the early stages [14]. Landholders with water supply bores in and around the CSG development areas were seeking better understanding of impacts and appropriate compensation, including provisions for alternative supply in advance of impacts occurring. There were also concerns about long-term impacts, monitoring and questions about the independence of scientific assessments presented in EISs by the CSG industry. At the same time, the industry was also seeking clarity on responsibilities where impacts may overlap.

3. Development of a groundwater management framework

To respond to the challenges outlined earlier, the existing regulatory arrangements were reformed to provide for a framework for independent assessment and management of cumulative impacts from petroleum and gas (P&G) development, including CSG. This was done in the context that, in Queensland, P&G tenure holders have a right to take groundwater that is unavoidably taken during the production of P&G. This right was made subject to certain obligations on managing impacts.

Development of the framework followed some key principles, such as: the cumulative assessment must be by an independent entity on a full-cost-recovery basis; the assessment and management arrangements must be periodic to adapt to evolution of knowledge of groundwater systems and changes to development; tenure holders must take the responsibility for making good the impacts on water supply bores; the management actions must be proactive; and there should be regular monitoring and reporting.

Two core elements of the framework that was established were: the establishment of an independent entity – the Office of Groundwater Impact Assessment (OGIA) – to displace proponents' responsibilities for preparing Underground Water Impact Reports (UWIR) in a declared Cumulative Management Area (CMA); and an ongoing three-yearly cycle of proponents preparing UWIRs containing impact assessment and management strategies, including the monitoring arrangements.

The entire intensive CSG development area, covering about 450 × 550 km, was declared a CMA (the Surat CMA) (**Figure 1**, Video 1). OGIA is tasked with assessing the cumulative groundwater impacts from P&G development in the Surat CMA. The costs associated with OGIA performing its functions are recovered through an annual levy payable by the tenure holders.

The technical assessment and arrangements for managing cumulative impacts are required to be reported by OGIA every three years in a UWIR. It is required that the report: includes an assessment and prediction of cumulative impacts in all affected aquifers; identifies the impact area for each aquifer; provides a list of affected water supply bores and GDEs; and outlines management arrangements such as the monitoring, make good of water supply bores, mitigation of impact on GDEs and assignment of responsibilities to tenure holders.

4. Implementation of the framework

The Surat CMA was declared in 2011. This was followed by the preparation and release of three iterations of assessments through UWIR 2012, UWIR 2016 and UWIR 2019.

The first cycle of assessment was completed within 12 months of the establishment of the CMA, primarily using the existing knowledge and secondary data sources to build a regional groundwater flow model and design management strategies. This was followed by a research program for the following three-year cycle, providing the foundation for the subsequent assessments in 2016 [2] and 2019 [3].

4.1 Hydrogeological assessment

A range of hydrogeological assessments were undertaken relating to the geology of the Surat Basin and aquifer interconnectivity. These investigations, in combination with complementary assessments by others, were then used to build a regional hydrogeological conceptualisation that underpinned the construction of a regional groundwater flow model and management strategies.

In the latest iteration, the geological model has 22 layers covering all major formations of the Surat and southern Bowen basins [15]. The geological model was based on the primary lithostratigraphic interpretation of geophysical logging from some 7,000 P&G wells. Primary interpretation ensures consistency in stratigraphic interpretation across the whole basin.

Most of the water use in the Surat Basin is for stock and domestic purposes, which is unmetered. Indirect estimates of water use were therefore made by developing a methodology utilising demand-based estimates per bore, while taking into account the availability of alternative water supply sources and seasonal variations [3]. The new methodology resulted in an estimate of groundwater use for stock and domestic purposes in the Surat Basin of 41,000 ML/year.

A major study was also undertaken to assess the connectivity of the overlying alluvial aquifer – the Condamine Alluvium [16]. The study involved multiple lines of investigation including drilling, coring, long-duration pump testing and monitoring. It concluded that there is a low level of connectivity.

A range of other complementary assessments was also undertaken including recharge estimation, fault characterisation and inter-aquifer connectivity. These studies provided the basis of a new regional conceptualisation [17].

4.2 Impact modelling

The most recent model developed by OGIA for the cumulative impact assessment in 2019 [15] represents the third iteration of conceptualisation, construction and calibration, based on information and data collected from monitoring and strategies developed in previous iterations. Each iteration of the model is informed by a revised understanding of key hydrogeological processes or concepts operating within the Surat CMA at the time.

The domain of the current model covers an area of around 460 × 650 km, encompassing the entire Surat CMA. The model domain is discretised into cells of 1.5 × 1.5 km areal extent, with 34 layers. The model is designed to simulate groundwater flow within the Surat Basin sequence and overlying alluvial formations in the Surat CMA, and within the CSG-producing Bandanna and Cattle Creek formations of the Bowen Basin.

The model was developed using the MODFLOW-USG simulator with a range of modifications to accommodate specific and unique processes associated with CSG extraction: the approximation of coal desaturation and dual-phase flow effects using a modified Richards equation formulation; use of a "descending drain" methodology to extract water from coal measures; recognition of the gas-filled status of CSG production wells and the consequential steep vertical gradient of water head in the vicinity of these wells; representation of 16 major fault systems in the groundwater model structure; and a new approach to parameterisation which maximised the use of the extensive lithological and other data from CSG well drilling activities.

Regional hydraulic properties were derived using numerical permeameters. The model was calibrated against a number of additional observation types including: monthly actual CSG extraction; vertical head differences between stratigraphic units; observed drawdowns; expected vertical head gradients; and saturations within the target formation.

The regional groundwater flow model was used to predict the impact of the cumulative industry development profile on groundwater pressures in aquifers. The profile was prepared based on information available at the time about historic and planned development of the individual CSG projects.

The current assessment [3] revealed that by the end of 2021, a total of 222 bores would be affected by a groundwater pressure reduction of more than five metres; these are referred to as Immediately Affected Area (IAA) bores. In the long term, a total of 571 water bores are predicted to be affected; these are referred to as Long-term Affected Area (LAA) bores.

4.3 Impact management

For all 222 water supply bores that are predicted to be impacted in the short term, i.e. IAA bores, follow-up actions are assigned to individual tenure holders – the responsible tenure holders (RTH) – based on certain rules. Each bore initially requires a bore assessment by the RTH to assess if the predicted impacts are likely to affect the intended purpose of the bore. If it is found that a bore water supply is likely to be impaired, then the RTH will have to reach a proactive 'make good' agreement with the bore owner.

The arrangement also involves the design and implementation of a monitoring network. Initially, in the first iteration, the UWIR 2012 specified the progressive installation of a network of 498 monitoring points across the Surat CMA. In recent iterations, the planned network has now been enhanced to 622 groundwater-level monitoring points and 103 water-quality monitoring points, of which about 500 are currently installed. The network is an extensive undertaking by tenure holders, considering the formations monitored are typically 200 to 1,000 m below ground. The UWIR assigned responsibilities to individual CSG companies for implementing individual parts of the regional monitoring network and reporting monitoring data.

The strategy for managing GDEs is primarily imbedded in a spring impact management strategy. Source aquifers for springs were established through investigations to predict impacts. Where the source aquifer for an identified spring was not known with confidence, the predicted pressure impact at the location of the spring was taken to be the maximum predicted pressure impact in any aquifer below the location of the spring.

4.4 Stakeholder consultation

OGIA undertakes formal and informal engagement activities to assist communities to understand the assessments that have been made, and to hear community views on groundwater impact issues. After the publication of a consultation draft UWIR, written submissions are invited and public meetings are held at community centres around the basin to hear questions and provide explanations, before finalising the UWIR. Current views on issues relating to groundwater impacts from CSG development can best be gauged by the submissions received on the consultation draft of the current UWIR 2019 [18].

Landholder and community groups have raised a range of issues relating to: the effect of both CSG and non-CSG groundwater take on the sustainability of the GAB, particularly in the Hutton Sandstone; the effect of climate change; the impact of migrating gas in water bores; delays in finalising 'make good' arrangements; the indirect impact of 'make good' bores in the Hutton Sandstone; overall impacts of CSG development; construction of CSG wells; the effect of the modelling scale on predicting impacts in water supply bores; and the inherent limitation associated with the modelling of impacts. There was a general expectation that, although many of the issues are outside the scope of the UWIR, broadening of scope should be considered in the future.

Issues raised during engagements are considered both in finalising the UWIR and in designing and implementing the subsequent research. For example, in 2012, the community raised specific concerns about connectivity between the target coal formation and the Condamine Alluvium. As a result, OGIA launched a research project on improving understanding of the connectivity through an extensive field program for data gathering and analysis. Ongoing community engagement on interim findings and field testing to build community understanding and confidence was an integral component of this program. Similar other engagements have continued, in collaboration with public and private sector organisations.

5. Conclusion

Extractive resource industries have a potential to impact groundwater resources. Particularly where the development is large-scale and involving multiple operators, the impacts can magnify due to their cumulative effects. In such situations, there are often a number of difficulties in managing impacts due to: different approaches to impact assessment by individual operators; lack of clarity on management responsibilities where impacts may overlap; constantly changing plans for development and evolving knowledge; and lack of community trust in assessments by industry.

These generic issues were well manifested in Queensland, Australia, where large-scale CSG development in 2010 brought them to the surface. In response, an innovative cumulative assessment and management framework applying adaptive groundwater management principles was developed and has been applied since then.

The framework involves an iterative cycle of independent impact assessment using progressively updated data and information, supported through secure funding arrangements. The cyclic assessment underpins progressive revision of strategies for managing impacts and enables identification of knowledge gaps to drive subsequent investigation.

The framework and its implementation are broadly regarded as effective in providing stakeholders with information and a mechanism to address issues relating to groundwater pressure impacts from CSG water extraction. As an independently funded and scientifically focused body, OGIA links assessment with regulatory management arrangements, and in doing so, has been able to build stakeholder confidence. For CSG companies, the framework provides clarity about statutory obligations. The involvement of OGIA reduces concerns about conflicts of interest, benefitting both CSG companies and bore owners.

Acknowledgements

TBD.

Videos

- Video 1: Introductory overview of the Surat CMA (https://www.business.qld.gov.au/industries/mining-energy-water/resources/environment-water/coal-seam-gas/surat-cma/location-geology).
- Video 2: The basins and geological formations, the complexity of the geological layers, effects of these layers, and movement of groundwater in the Surat CMA (https://www.business.qld.gov.au/industries/mining-energy-water/resources/environment-water/coal-seam-gas/surat-cma/location-geology).
- Video 3: How groundwater impacts may occur in aquifers surrounding the CSG formations in the Surat Basin. (https://www.business.qld.gov.au/industries/mining-energy-water/resources/environment-water/coal-seam-gas/surat-cma/location-geology).

Author details

Sanjeev Pandey[1*], Randall Cox[2] and Steven Flook[1]

1 Office of Groundwater Impact Assessment, Brisbane, Australia

2 Brisbane, Australia

*Address all correspondence to: sanjeev.pandey@rdmw.qld.gov.au

References

[1] Queensland Government (2017) Petroleum and gas production and reserve statistics. https://data.qld.gov.au/dataset/petroleum-gas-production-and-reserve-statistics

[2] OGIA (2016a) Underground Water Impact Report for the Surat Cumulative Management Area. OGIA, Department of Natural Resources and Mines, Brisbane

[3] OGIA (2019a) Underground Water Impact Report for the Surat Cumulative Management Area. https://www.dnrme.qld.gov.au/__data/assets/pdf_file/0003/1445547/uwir-2019.pdf

[4] Cooper LM, Sheate WR (2002) Cumulative effects assessment: A review of UK environmental impact statements. Environ Impact Assess Rev 22:415-439. doi: 10.1016/S0195-9255(02)00010-0

[5] Rahm BG, Riha SJ (2012) Toward strategic management of shale gas development: Regional, collective impacts on water resources. Environ Sci Policy 17:12-23. doi: 10.1016/j.envsci.2011.12.004

[6] Bonnell S, Storey K (2000) Addressing cumulative effects through strategic environmental assessment: A case study of small hydro development in Newfoundland, Canada. J Environ Assess Policy Manag 2:477-499

[7] Gunn J, Noble BF (2009) Integrating cumulative effects in regional strategic environmental assessment frameworks: Lessons from practice. J Environ Assess Policy Manag 11:267-290. doi: 10.1142/S1464333209003361

[8] Noble (2008) Strategic approaches to regional cumulative effects assessment: a case study of the Great Sand Hills, Canada. Impact Assess Proj Apprais 26:78-90

[9] Therivel R, Ross B (2007) Cumulative effects assessment: Does scale matter? Environ Impact Assess Rev 27:365-385. doi: 10.1016/j.eiar.2007.02.001

[10] Johnson D, Lalonde K, McEachern M, et al (2011) Improving cumulative effects assessment in Alberta: Regional strategic assessment. Environ Impact Assess Rev 31:481-483. doi: 10.1016/j.eiar.2011.01.010

[11] Hodgkinson J, Preda M, Hortle A, et al (2009) The potential impact of carbon dioxide injection on freshwater aquifers: The Surat and Eromanga Basins in Queensland. Department of Employment, Economic Development and Innovation, Brisbane

[12] Queensland Water Commission (2012) Underground water impact report for the Surat Cumulative Management Area. Queensland Water Commission, Brisbane

[13] Office of the Chief Economist (2015) Review of the socioeconomic impacts of coal seam gas in Queensland. http://www.appea.com.au/wp-content/uploads/2015/10/Socioeconomic-impacts-of-CSG-in-Queensland.pdf

[14] Gillespie N, Bond C, Downs V, Staggs J (2016) Stakeholder trust in the Queensland CSG industry. APPEA J 56:239-246

[15] OGIA (2019b) Updated Geology and Geological Model for the Surat Cumulative Management Area. Department of Natural Resources, Mines and Energy, Brisbane

[16] OGIA (2016b) Groundwater Connectivity Between the Condamine Alluvium and the Walloon Coal Measures. OGIA, Department of Natural Resources and Mines, Brisbane

[17] OGIA (2016c) Hydrogeological Conceptualisation Report for the Surat Cumulative Management Area. OGIA, Department of Natural Resources and Mines, Brisbane

[18] OGIA (2019c) Submissions Summary: Consultation on the Draft Underground Water Impact Report 2019 for the Surat Cumulative Management Area. https://www.dnrme.qld.gov.au/__data/assets/pdf_file/0006/1461237/uwir-submissions-summary.pdf

Section 3

Water Resources

Chapter 5

Sources Identification of Water Inrush in Coal Mines Using Technique of Multiple Non-Linear Machine Learning Modelling

Yao Shan

Abstract

Water inrush is a major threat to the working safety for coal mines in the Northern China coal district. The inrush pattern, threaten level, and also the geochemical characteristics varies according to the different of water sources. Therefore, identifying the water source correctly is an important task to predict and control the water inrush accidents. In this chapter, the algorithms and attempts to identify the water inrush sources, especially in the Northern China coal mine district, are reviewed. The geochemical and machine learning algorithms are two main methods to identify the water inrush sources. Four main steps need to apply, namely data processing, feature selection, model training, and evaluation, in the process of machine learning (ML) modelling. According to a calculation instance, most of the major ions, and some trace elements, such as Ti, Sr, and Zn, were identified to be important in light of geochemical analysis and machine learning modelling. The ML algorithms, such as random forest (RF), support vector machine (SVM), Logistica regression (LR) perform well in the source identification of coal mine water inrush.

Keywords: water inrush, source identification, coal mines, non-linear machine learning, groundwater

1. Introduction

Water inrush is one of severe hazards to coal mines in China. According to statistical material, more than 25 billion tons of coal resources are at the risk of water inrushes in China. From 2000 to 2015, 1162 water inrush accidents were reported, causing 4676 deaths. The number of accidents and deaths took 3.3% and 7.8% of all accidents in coal mines. In spite of the low proportion, major accidents often took place, leading to severe property and live loss.

Northern China district is an important coal base area, reserves of which takes nearly 40% of all country. Therefore, the prevention of water inrush accident is a key issue to the mining safety. The main threats of water inrush to the working face can be grouped into mainly four types, namely surface water, coal roof aquifer water, coal floor aquifer water, and goaf water. The coal roof water is usually relative to coal seam sandstone aquifers, sometime associate with quaternary aquifers. Goaf water formed when the working face closed and ground water filling up this space. The coal floor water is usually relative to limestone aquifers in the Ordovician system and the Taiyuan Formation in the Carboniferous system.

The different types of water inrush threats show various foreshadow, bursting behaviour, and hazard rating, and corresponding treating technology is essential. Therefore, the technique to predict and evaluate the accident potential, forecast the accident occurrence, and identify water inrush sources, is a key step to prevent the accidents or disasters, and protect the working safety and human health and lives.

In this chapter, the main techniques that used to identify water inrush sources and its application, mainly focusing on the Northern China district, are illustrated.

2. Methods and it's applications for source identification

A basic strategy to identify source of water inrush is based on the geochemical characteristics. Some researchers have compared concentrations of major ions, including K^+, Na^+, Ca^{2+}, Mg^{2+}, Cl^-, SO_4^{2-}, CO_3^{2-}, HCO_3^-, and also total dissolved solid, between different aquifers to determine water sources.

In different aquifers, the water composition is a response of its original characteristics and water-rock interaction process. In the Northern China, two main groups of aquifers are coal bearing strata aquifers and limestone aquifers. The geochemists used to find key ions in water, sometimes using geochemical figures, to determine the water sources.

While the geochemical strategy is based on some unique ions and parameters in a lower dimension, another strategy, namely the machine learning (ML) algorithms, is based on multivariate analysis, including some specific methods, and provide more quantitate and reliable results.

2.1 Geochemical methods

The geochemical method is a popular technique in the water inrush identification, for mainly two reasons. First, some coal mines, especially for the large companies, have their own laboratory to test water geochemistry. Therefore, it is easy to obtain data. Second, the experienced technicians are familiar with the water geochemical data, especially for the major ions and important parameters. Researchers usually begin their study from the normal water geochemistry, to investigate water characteristics in every aquifer, set up identification model to distinguish water type from others, and find out the water-rock interaction mechanism for the water composition.

An easy-to-handle method to identify water source is to analyse the major ion characteristics. Cheng et al. [1] analysed water geochemistry in quaternary aquifers, magmatic aquifers, limestone aquifers in the Huaibei coal mine district, Anhui province. The data was grouped into different chemical types, which can be used as database for the water source identification. Chen and Gui [2] discussed water geochemistry in Wanbei coal mine district in Anhui province. Zhang and Cao [3] analysed ground water in Hancheng coal mine district in Shannxi province, founding that the potential water burst point was related with the limestone aquifer. Dai et al. [4] discussed water characteristics in Xiangshan coal mine in Shannxi province. The data was grouped using SPSS to set up a database for further coal mine monitoring and forecasting. The author's group have collected and analysed more than 30 water samples in the Lu'an coal mine district in Shanxi province, the pattern of water flow underground and water characteristics in every ground aquifer were summarised, some important ions, trace elements, and parameters, were identified and used to distinguish water sources from others.

A geochemical chart, the piper diagram, is usually used to analyse and group water samples into different groups by drawing the data as points in two triangle and a diamond figure. Zhang and Cao [3], Dai et al. [4] have applied this technique to identify the water sources. Author's group have collected samples in 2019, **Table 1** shows part of the data, and **Figure 1** shows the water geochemistry in a piper diagram.

As **Figure 1** showing, water in coal bearing seam shows similar characteristics, Na^+ and K^+ take more than 80% and up to more than 95% of all the cations. TDS of the most water sample were less than 1000 mg. The limestone water shows a spanning pattern. TDS of the limestone aquifer water also showed a much wider range, from less than 500 mg to higher than 3000 mg. In the limestone aquifer, water volume is larger, and water-rock interaction is stronger than that in the coal bearing seam, which maybe the reason to the water characteristics in the limestone aquifers.

The source identification using basic geochemical technique is a qualitative, or semi-quantitative method, which may mainly depend on researchers' experiences. If distinguished differences between aquifers are observed in low dimensions, the basic geochemical technique is useful and easy to use. However, while the differences reveal in a higher dimension, i.e. the difference of ions' composition, this method system may lead to a confusing result.

Not only the major ions, but also trace element concentrations and isotope values can be used to distinguish one source from others. Some researchers have used the trace elements to distinguish water samples from others or set up discriminant models. Feng and Han [5] analysed concentration and occurrence of trace elements and modelled its formation using PHREEQC. Chen et al. [6] collected 24 samples from the quaternary aquifers, coal seam sandstone aquifers, and limestone aquifers in Wanbei coal mine district in Anhui province and tested 24 types of trace elements, including Be, B, Sc, V, Cr, etc. The samples and trace elements were clustered. Then eight trace elements, including Be, Zn, Ga, Sr, U, Zr, Cs, Ba, were found to be key parameters to set up discriminant model. The key trace elements were used to train Bayes discriminant analytical model with a good performance.

Isotopes are also used in the water inrush in the coal mines. The most popular isotopes are δD and $\delta^{34}S$ of water. In recent years, the studies are applied in Wanbei coal mine district [7–9], and Fushun coal mine district [10], etc.

No.	K^++Na^+	Ca^{2+}	Mg^{2+}	Fe^{3+}	Fe^{2+}	NH_4^+	Total
1	363.95	0.84	0.23	0.03	0.09	0.00	365.14
2	17.07	138.56	38.96	0	0	0	194.59
3	12.86	61.84	13.63	0	0	0	88.33
4	41.58	69.88	32.14	0	0	0.07	143.67
5	12.01	65.87	32.14	0	0	0.2	110.22
6	240.88	0.08	0.02	0.15	0	0.44	241.57
7	315.05	12.44	4.14	0	0	1.51	333.14
8	88.16	356.63	38.96	0.11	0	0.38	484.24
9	351.19	1.07	0.54	0.07	0	0.47	353.34
10	334.4	0.21	0.21	0	0	0.41	335.23

Table 1.
Major ion data in ground aquifer (mg/L).

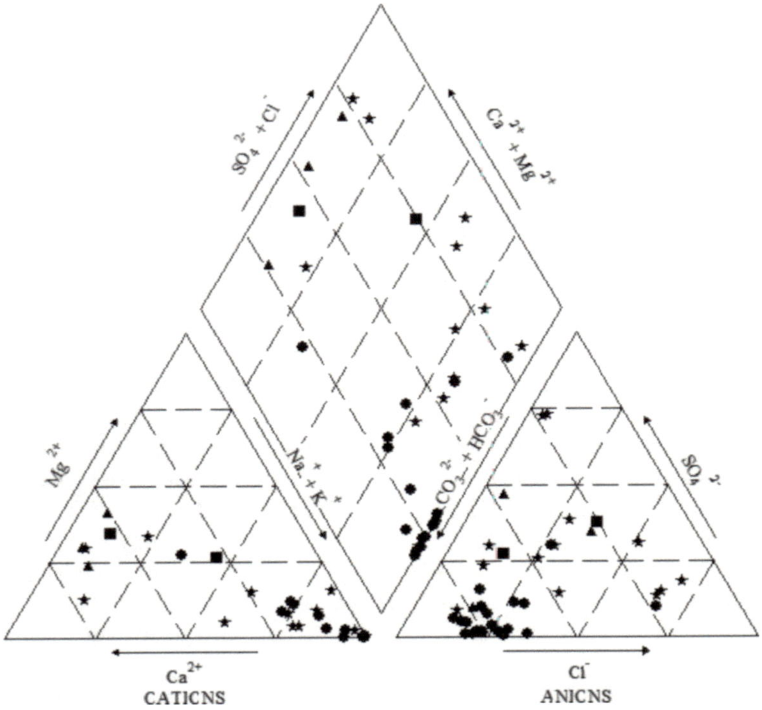

Figure 1.
Piper drawing of the ground water (In the figure, the squares stand for surface water, the triangles stand for quaternary water, the circles stand for coal bearing seam water, and the stars stand for limestone aquifer water).

In the author's research in Lu'an coal mine district in Shanxi province, the major ions and trace element were treated together, then SO_4^{2-}, Ti, Sr, Mg, K + Na, Zn, and Cl^- were chosen to be typical ions or elements to train models.

Furthermore, the water form in a scale of whole water unit, therefore the analysis should be carried out in a scale of whole water unit, but not a single point. In the Northern China area, several ground units can be divided, the water-rock interactions among which show similar pattern in different coal mine district. Therefore, the analysis of coal mine district scale and comparison between different coal mine district is an important task to summarise the common mechanism of the water-rock interaction and distinction models.

2.2 Machine learning methods

The geochemical method is effective only if the water samples can be grouped and divided very clearly by one or very few parameters. In most scenario, the ion-distinguishing method is confusing and lack of accuracy. The difference of water samples is embedded in a high dimension, i.e. the combination of major ions, trace elements and other parameters. It is hard to find the dividing mode just by observation or simple drawing. Benefiting from the developing of data science and technology, the environmental and geological issues, including the ground water can be described, and divided by ML methods.

The ML algorithm can be simply divided into supervised, unsupervised, and semi-supervised, depending on how the target variables are labelled. For some environmental and geological problems, the target variables cannot be labelled, then the unsupervised ML algorithm, such as principal components analysis (PCA) are applied. For example, Shan et al. [11] applied the PCA method to analyse the occurrence and leaching mechanism in coal and host rock, Pumure et al. [12] found out successfully of the occurrence of As and Se in coal host rock. Self-Organising Maps (SOM) is a kind of unsupervised artificial neural network (ANN) used in a large data amount scenario [13–14]. The PCA algorithm is only used for the water inrush if the target variables cannot be labelled [15–17].

While the researchers carrying out their studies, discriminant models should be trained. The training data is obtained from the samples collected from every aquifer. In this step, the data is usually marked clearly. Therefore, the target variables can be obtained for most research cases, and the supervised ML algorithm can be used, which shows high precise and accuracy than the unsupervised ML algorithm. There are several algorithms are suitable for the model training, such as artificial neural network (ANN), support vector machine (SVM), discriminant analysis (DA), decision tree (DT), random forest (RF), boosting, and regression, etc.

In the Northern China, supervised ML algorithm has been used in several coal mine districts. **Table 2** shows part of the research cases in the Northern China area in recent years. It can be concluded from the table that DT criterions are most implemented, some other methods, such as SVM, and ANN, are also used.

2.2.1 Supervised machine learning algorithm

Up to present, the DA is a most popular method analysis to identify sources of water inrush in the Northern China district. Two criterions are usually used, namely Fisher criterion and Bayes criterion. In the framework of Fisher-criterion based DA algorithm, high dimensional data is projected to a one-dimension space, then a discriminant criterion is obtained to achieve the maximum variance between two groups and the minimum in-group variance. Because this method is used to handle a two-group problem, many rounds of calculation are needed for a multiple-group problem. The Bayes-criterion base DA method calculate the posterior probabilities of the sample in each group, then the sample can be classified into the group with the highest posterior probability. Comparing with the Fisher criterion, the Bayes criterion is more frequently used.

The DA is a kind of linear algorithm. Along with the development of ML technology, non-linear modelling is widely used in researches, including the geological, environmental, and engineering area. In order to deal with problems of surface water and ground water, the SVM method is applied to predict water quality and water level [23, 24], ANN and DT are used to predict the [NO_3^-] of ground water [25], set up the water quality monitoring system [26]. Boosting tree is also used to classify distributed water and ground water.

However, the non-linear ML method is relatively less applied to deal with the water inrush problems in coal mines, though higher accuracy maybe achieved

Coal mine districts	Years	Algorithm	Sources
Anhui province	2014	Fishes discriminant	[18]
Henan Province	2015	Feature ions- discriminant	[19]
Shanxi Province	2015	Support vector machine	[20]
Anhui Province	2016	Bayes Discriminant	[6, 7, 21]
Henan province	2019	PCA – grey relationship analysis	[22]

Table 2.
Some implementation of water inrush source identification in the northern China in recent years.

compared to the linear algorithm. According to literature research, the ANN [27] and SVM [20] have been implement in this area. The ANN is a very popular technique in many areas, including figure and voice identification, driverless driving, etc. However, the ANN usually needs large amount of data to train model to control its over-fitting problem. On the other hand, the data of the environmental and geological area, including the water inrush analysis are usually structured data, and limited to a small data quantity. As a result, the problem of over-fitting problem is hard to control, which means low accuracy of prediction is prospected while using the ANN model to check using the testing data, though a high accuracy may be obtained while testing the model using the training data. The algorithm of SVM perform better to control the over-fitting problem. Other than SVM, the DT, DT, boosting tree, Bayes network (BN) also have good prospect, in consideration of the characteristics of the coal mine ground water, i.e. structured small data quantity.

2.2.2 Data selection and feature engineering

The tested data of ground water is material of model training. However, the data preparing is essential to ensure or enhance the model quality. The data preparing work mainly includes data selection and feature engineering.

In a wide sense, the data selection includes data cleaning, which means treatment of unit and missing data. Then the data should be selected to determine those used in the model training step. The data selection is applied in two stages, before model training and after model training. Before the model training, the suitable data for the model training means to make sure all the data is labelled correct. Uncorrected marked data leads to wrong model definitely, regardless of the quality of models. After the model training, the training data should be checked again. The data have to be checked very carefully to find wrong classified data. While it is determined to wrong pre-labelled, then the data should be deleted, and new model need to be trained.

The other important work before the model training is feature engineering. The basic mechanism to process feature engineering is to achieve a best performance of the model. **Figure 2** shows the idea of feature engineering. Number of features, or parameters, means the model complexity. More features in the model lead to a higher complexity of the model. As **Figure 2** showing, the prediction performance is related to model complexity. A very simple model lead to very bad model

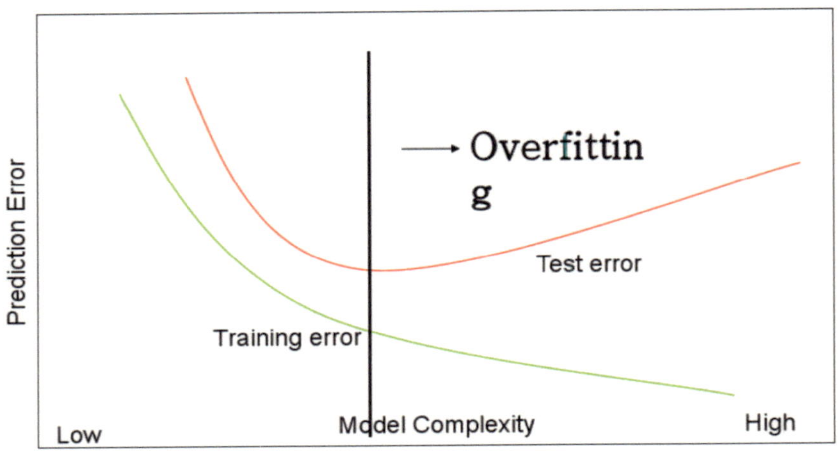

Figure 2.
Correlation of the model complexity and prediction error.

performance, that's why the non-linear models are used. Along with the increasing of model complexity, the prediction error of the training samples becomes lower steadily. On the other hand, the prediction error of the testing samples gets lower at first, then higher again. That suggested over-fitting problem in the ML model. Therefore, the feature has to be processed if a good performing model is acquired.

The feature engineering includes feature fusion and feature selection. A common feature fusion method is PCA. The PCA can reduce dimension of data, then the features in a lower space could stand for most data information. As combination of the original features, the new feature cannot reflect the data characteristics directly. While the researchers want to analyse the importance of the parameters in the original data, the feature selection technique should be used.

Popular feature methods include RF, and Lasso regression, etc. The RF based feature selection undergoes the following steps.

1. The data set X contain N samples, draw samples randomly from the data set X using the bootstrap resampling method. The resampling is carried out k times, to construct k regression tree. In this process, the probability of no drawing of each sample is $p = (1-1/N)^{N}$. The p tends to 0.37 while the N increasing to infinity. That means that about 37% of the samples in the data set X are not drawn, these data are not used in the DT training, calling out-of-bag (OOB) data. These OOB data is used to test the regression trees.

2. For k bootstrap samples, k unpruned regression trees are created respectively. In the training process, for each node, m attributes are randomly selected from the total M attributes as internal nodes. Then, an optimal attribute is selected from m attributes as a split variable to make the branches grow, according to the minimum Gini index principle.

3. The k decision tress comprises a random forest, the model quality could be evaluate using two indices: large mean square error of OOB (MSE_{OOB}) and low coefficients of determination (R^2_{RF}).

$$MSE_{OOB} = \frac{\sum_{i=1}^{n}(y_i - \hat{y}_i)^2}{n} \tag{1}$$

$$R^2_{RF} = 1 - \frac{MSE_{OOB}}{\hat{\sigma}_y^2} \tag{2}$$

Where n is the total number of the samples, $\hat{y}_i$ is the predicted output obtained by the generated RFR regression model, y_i is the observed output value, and the $\hat{\sigma}_y^2$ is the predicted variance of the OOB output.

4. The RF regression model provides two methods to determine the importance degree of each variable index: mean decrease in Gini index and mean decrease in accuracy. In a regression model, the mean decrease in Gini is usually used, and the mean decrease in accuracy is more applied for the classification problem. The water inrush source identification is a kind of classification problem, therefore the mean decrease in accuracy is selected.

While carrying out the inrush source identification, the attributes could be used in the model includes major ions, trace elements, important parameters, and isotopes, etc. In which, the data of major ions and important parameters are easier to obtain. On the other hand, adding of trace elements and isotopes into the models may enhance the model performance, for these parameters carries a lot of information of the water samples. In consideration of easy using, only major ions and important parameters is used, while considering for the model accuracy, more parameters could be added. Therefore, it is a balance need to consider while building models.

In our previous study in the Lu'an coal mine district in Shanxi Province, all the prescribed parameters have been tested. In the first step, the feature selection was applied on the major ions and important parameters. **Figure 3** shows the calculation result while using the algorithm of RF. As **Figure 3** showing, key attributes to the mode are SO_4^{2-}, C (stands for CO_3^{2-} and HCO_3^-), K^++Na^+, Mg^{2+}, Cl^-, and Ca^{2+}, in a

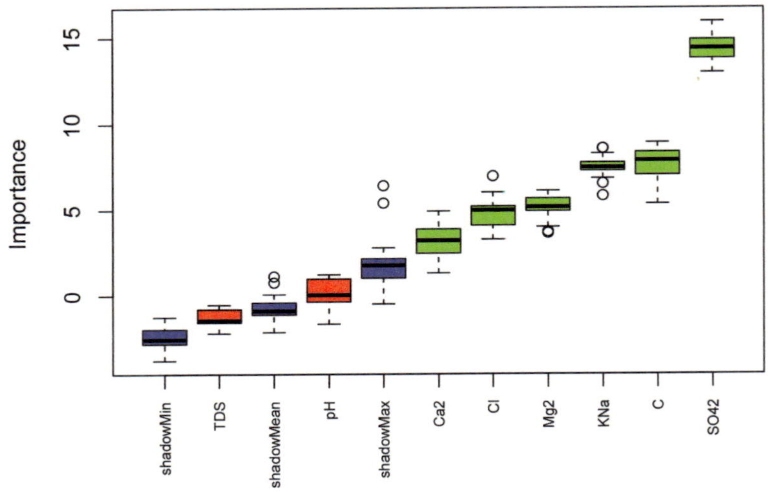

Figure 3.
Feature selection result of major ions and important paraments using RF algorithm.

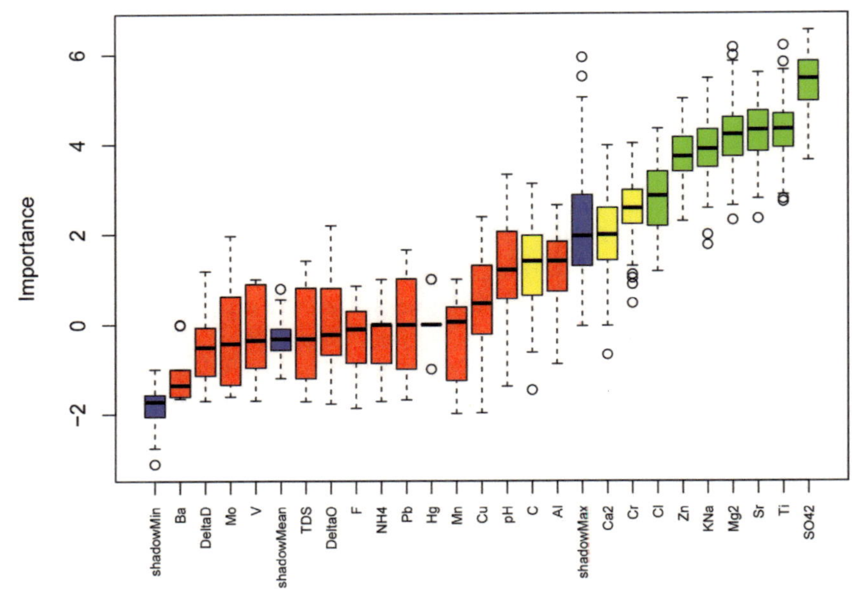

Figure 4.
Feature selection result of all data using RF algorithm.

descending order. It can be concluded that all the major parameters contribute to the explanation of water source identification.

For the trace elements and isotopes may carry information of the water samples, all the parameters were calculated following the same process of feature selection illustrated in the previous section, which is shown in **Figure 4**. As the figure showing, the key attributes for the model turned out to be SO_4^{2-}, Ti, Sr, Mg^{2+}, K^++Na^+, Zn, and Cl^-. Comparing with the first feature select, the major ions still play important role in the water inrush identification, that's because of the significant water-rock interaction in the ground aquifer. However, some trace elements are also important in the distinguishing water source from others, including Ti, Sr, and Zn. These trace elements show low concentrations in coal bearing seam water, and higher concentrations in limestone aquifer water.

We have also applied the Lasso regression to determine important attributes for the model training. The RF perform better, so only the RF result is illustrated in this chapter.

2.2.3 Model selection

In the previous description, the data and attributes have been selected to train models, and the model frameworks need to select and evaluate. According to the literature review, the non-linear ML modelling is seldom applied to solve the problem of water inrush in coal mines in the Northern China area. However, by consideration the mechanism of ML algorithm, and relative studies in the environmental and geological area, some non-linear ML method were applied, evaluated and compared.

Li et al. [20] applied the SVM algorithm to determine the source of water inrush. The SVM algorithm project data from a lower dimensional space to a higher dimensional space by applying the kernel functions. Then the data that cannot divided in a lower dimensional space can be separated. Several types of kernel can be evaluated, such as radial basis and poly-nominal functions, etc.

Li et al. [20] compared the performance of geochemical and machine leaning methods on the water inrush issues based on the Lu'an coal mine district. It was found that, the accuracy by using the SVM method has achieved 100%, while only 42% and 48% have obtained while using one (SO_4^{2-}) or two (SO_4^{2-} and Ca^{2+}) parameters as the key attributes, respectively. In this research, the radial basis and poly-nominal function were not compared. However, a mixing kernel function was proposed.

The ANN has also used to identify the source of coal mine water inrush [27]. For example, in the research of Li et al. [27], an improved genetic algorithm combining PCA and back propagation ANN were applied. The research had achieved a simple, reasonable, and effective distinguishing result.

In a previous study carried out by the authors' group, 42 samples were collected from the Lu'an coal mine district, Shanxi province, which belong to six water types, namely sandstone aquifer water, Ordovician limestone water, Taiyuan formation limestone water, goaf water, spring water, and fault-oxidising zone water, respectively. The parameters tested included K^+, Ca^{2+}, Mg^{2+}, SO_4^{2-}, pH, Fe, NO_2^-, I, Cl^- and temperature. The data was normalised before the model training. Some non-linear ML was applied. The research has focused on the following issues: first, the key attributes should be selected by the ML method, which has been described in the previous section, second, the data need to be cleaned and selected, third, a wide scope of parameters should be considered, including major ions, important traditional parameters, trace elements, and isotopes, and the fourth, the ML framework need to be evaluated and compared.

We have applied several algorithms; the identification result is shown in **Table 3**. Considering the easy using of the model, only major ions, and key parameters are used

Attribute scope	ML algorithm	Corrected identification
Major ions and parameters	RR	28/31
Major ions and parameters	RF	29/31
Major ions and parameters	SVM (radial basis)	29/31
Major ions and parameters	SVM (poly-nominal)	30/31
Major ions and parameters	LR	29/31
Major ions and parameters	DA	29/31
All data	RR	23/25
All data	RF	25/25
All data	SVM (radial basis)	24/25
All data	SVM (poly-nominal)	25/25
All data	LR	25/25
All data	DA	25/25

Table 3.
Identification result of water resource using the non-linear ML algorithm.

in the first-round model training. In a traditional process, the data should be divided into two groups with a ratio of 7:3. The first part is used to train models, and the second part is used to evaluate the model performance. Because the amount of data is limit, the models' test was carried out by using the method of re-calculating the sample data using models. Other than the ridge regression (RR), other algorithm showed better performance. The DA was based on the Bayes criterion, showing an acceptable result. Comparing the two SVM method, poly-nominal kernel got a better result than that of the radial kernel SVM.

The wrong identified data is important to the modelling. Four main reason may lead to the result: First, the water sample was wrong labelled; second, the water is a mix of multi water sources; third, the attributes selected is not suitable; and fourth, the model need to be improved or more suitable model framework is needed.

For the first reason, the data should be analysed very carefully. If not representative, the samples should be deleted. For the second reason, the water mix should be calculated using other method, such as the nonlinear programming. In order to improve the model performance, the scope of the attributes enlarged to include trace elements and isotopes. For the trace elements of some samples was not tested, the overall sample data was less than the first-round training. According to the results, better performance was observed when using the trace elements. It suggested that some trace elements can be used as the key attributes in the water source identification. And the wrong-identified samples were probably because of poorer performance of the first-round modelling, rather than the wrong-labelling of the samples. Comparing the ML framework, the RF, SVM, LR, and the Bayes-based discriminant analysis all showed high accuracy. The accuracy of ridge regression (RR) is less than others. Comparing the two kernels used for the SVM, the poly nominal based SVM showed a better performance.

3. Conclusions

In this chapter, the main methods to identify water inrush sources in the coal mines, especially in the Northern China area, were reviewed.

The basic idea of the geochemical method is to find characteristic ions, which shows distinct behaviour in one group of water samples from others, by geochemical analysis and drawing.

In the process of non-linear machine learning, four main steps need to apply, they are: data processing, feature selection, model training, and evaluation. According to the studies of literature review and the author's previous study, some key attributes/ parameters were selected using the ML algorithm, then the ML algorithm was compared. It was found that, most of the major ions, and some trace elements, such as Ti, Sr, and Zn, were important in the model training. Then ML algorithms, RF, SVM, LR perform well in the source identification of coal mine water inrush.

Acknowledgements

Our study is funded by the Fundamental Research Funds for the Central Universities (3142014005) and the Colleges and Universities in Hebei Province Science and Technology Research Project (ZD2016204).

Conflict of interest

The authors declare no conflict of interest.

Author details

Yao Shan
School of Emergency Technology and Management, North China Institute of Science and Technology, Yanjiao, 101601, China

*Address all correspondence to: 9106350@qq.com

References

[1] Cheng C, Ge X, Wang, D. Hydrogeochemical characteristics of Baishan coal mining area and the application to the source discrimination of water invasion in the mine. Journal of Huainan Mining Institute. 1994; 14(1): 25-32. (In Chinese)

[2] Chen L, Gui H, Hu Y, Xu G, Yuan W. Study of hadrochemical characteristics in Karst water of coal seam floor in the Mining District of North Anhui. Coal Geology & Exploration. 2003; 31(2): 27-31. (In Chinese)

[3] Zhang L, Cao H. Distinguishing the sources of water inrush in Sangshuping coal mine by hadrochemical characteristics. Coal Geology & Exploration. 2013; 41(4): 42-45. (In Chinese)

[4] Dai G, Xue X, Niu C. Discrimination of water inrush source in xiangshan coal mine based on chemical characteristics of groundwater. Journal of Xi'An University of Science and Technology. 2017; 37(2): 213-218. (In Chinese)

[5] Feng Q, Han B. Occurrence of trace elements in reservoir, oilfield water, and scaling of Wumishan Formation, Renqiu, China. Journal of China University of Mining & Technology. 2001; 30(2): 122-125. (In Chinese)

[6] Chen L, Gui H, Yin X, Qian J, Pan L. The standard type trace elements and the discriminant model of water bursting source in the Linhuan coal district. Hydrogeology & Engineering Geology. 2010; 37(3): 18-22. (In Chinese)

[7] Gui H, Chen S. sotopic geochemical characteristics of groundwater and its geological significance in Sunan mining area. Earth Science Frontiers. 2016; 23(3): 133-139. (In Chinese)

[8] Chen L, Gui H, Yin X. Composing characteristic of hydrogen and oxygen stable isotopes and tracing of hydrological cycle. Journal of China Coal Society. 2008; 33(10): 1107-1111. (In Chinese)

[9] Chen L, Gui H, Yin X, Qian J. Characteristic 0of 18 O and D and Current Field in Deep Groundwater. Journal of China University of Mining & Technology. 2008; 37(6): 854-859. (In Chinese)

[10] Hu W, Ma Z, Cao H, Liu F, Li T, Dou H. Application of isotope and hydrogeochemical methods in distinguishing mine bursting water source. Journal of Earth Sciences and Environment. 2010; 32(3): 268-271. (In Chinese)

[11] Shan Y, Wang W, Qin Y, Gao L. Multivariate analysis of trace elements leaching from coal and host rock. Groundw Sustain Dev. 2019; 8: 402-12.

[12] Pumure I, Renton JJ, Smart RB. The interstitial location of selenium and arsenic in rocks associated with coal mining using ultrasound extractions and principal component analysis (PCA). J Hazard Mater. 2011;198:151-8.

[13] Lacassie JP, Roser B, Ruiz Del Solar J, et al. Discovering geochemical patterns using self- organizing neural networks: a new perspective for sedimentary provenance analysis. Sedimentary Geology. 2004; 165(1-2): 175-191.

[14] Nakagawa K, Yu Z-Q, Berndtsson R, et al. Temporal characteristics of groundwater chemistry affected by the 2016 Kumamoto earthquake using self-organizing maps. Journal of Hydrology. 2020; 582: 1-14.

[15] Valder J F, Long A J, Davis A D, et al. Multivariate statistical approach to estimate mixing proportions for unknown end members. Journal of Hydrology. 2012; 460-461: 65-76.

[16] Menció A, Folch A, Mas-Pla J. Identifying key parameters to differentiate groundwater flow systems using multifactorial analysis. Journal of Hydrology. 2012; 472-473: 301-313.

[17] Ju Q, Hu Y, Zhang S. Mine water inrush source identification method based on principal component analysis and Bayesian discriminant. Coal Engineering. 2018; 50(12): 90-94. (In Chinese)

[18] Yang X, Wang L, Zhang Z. The Application of Fisher Approach to Discrimination of Mine Water-inrush Source. Science & Technology Information. 2014; 11: 97-98. (In Chinese)

[19] Zhu Q, Li X. Water source discrimination of inrush water in mine based on water chemistry characteristics analysis. Journal of North China Institute of Science and Technology. 2015; 12(3): 29-33. (In Chinese)

[20] Li J, Xu Y, et al. Research on method of automatic recognition of water sources based on Support vector machine. Journal of North China Institute of Science and Technology. 2015; 12(2): 25-29. (In Chinese)

[21] Chen M, Wu Y, Gao D, Chang M. Shallow Groundwater Hydrogeochemical Evolution Process and Controlling Factors in Plain Zone of Guanghan City. Journal of Jilin University (Earth Science Edition). 2016; 46(3): 831-843. (In Chinese)

[22] Yang Z, Ding F, Huang P. Research on PCA-GRA Discrimination Model of Water Inrush Source in Mines. Coal Technology. 2019; 38(12): 84-87. (In Chinese)

[23] Singh K P, Basant N, Gupta S. Support vector machines in water quality management. Analytica Chimica Acta. 2011; 703(2): 152-162.

[24] Kisi O, Shiri J, Karimi S, et al. A survey of water level fluctuation predicting in Urmia Lake using support vector machine with firefly algorithm. Applied Mathematics and Computation. 2015; 270: 731-743.

[25] Wagh V, Panaskar D, Muley A, et al. Neural network modelling for nitrate concentration in groundwater of Kadava River basin, Nashik, Maharashtra, India. Groundwater for Sustainable Development. 2018; 7: 436-445.

[26] O'Connor E, Smeaton A F, O'Connor N E, et al. A Neural Network Approach to Smarter Sensor Networks for Water Quality Monitoring. Sensors. 2012; 12(4): 4605-4632.

[27] Li Y, Niu G, Liu H. Application of improved GA-BP neural network on identification of water inrush source in mine. Journal of Safety Science and Technology. 2016; 12(7): 77-81. (In Chinese)

Chapter 6

Fractal Scaling Properties in Rainfall Time Series: A Case of Thiruvallur District, Tamil Nadu, India

Ibrahim Lawal Kane and Venkatesan Madha Suresh

Abstract

In the present study, the features of rainfall time series (1971–2016) in 9 meteorological regions of Thiruvallur, Tamil Nadu, India that comprises Thiruvallur, Korattur_Dam, Ponneri, Poondi, Red Hills, Sholingur, Thamaraipakkam, Thiruvottiyur and Vallur Anicut were studied. The evaluation of rainfall time series is one of the approaches for efficient hydrological structure design. Characterising and identifying patterns is one of the main objectives of time series analysis. Rainfall is a complex phenomenon, and the temporal variation of this natural phenomenon has been difficult to characterise and quantify due to its randomness. Such dynamical behaviours are present in multiple domains and it is therefore essential to have tools to model them. To solve this problem, fractal analysis based on Detrended Fluctuation Analysis (DFA) and Rescaled Range (R/S) analysis were employed. The fractal analysis produces estimates of the magnitude of detrended fluctuations at different scales (window sizes) of a time series and assesses the scaling relationship between estimates and time scales. The DFA and (R/S) gives an estimate known as Hurst exponent (H) that assumes self-similarity in the time series. The results of H exponent reveals typical behaviours shown by all the rainfall time series, Thiruvallur and Sholingur rainfall region have H exponent values within $0.5 < H < 1$ which is an indication of persistent behaviour or long memory. In this case, a future data point is likely to be followed by a data point preceding it; Ponneri and Poondi have conflicting results based on the two methods, however, their H values are approximately 0.5 showing random walk behaviour in which there is no correlation between any part and a future. Thamaraipakkam, Thiruvottiyur, Vallur Anicut, Korattur Dam and Red Hills have H values less than 0.5 indicating a property called anti-persistent in which an increase will tend to be followed by a decrease or vice versa. Taking into consideration of such features in modelling, rainfall time series could be an exhaustive rainfall model. Finding appropriate models to estimate and predict future rainfalls is the core idea of this study for future research.

Keywords: Hurst Exponent, Detrended Fluctuation Analysis (DFA), Rescaled Range (R/S) Method, Fractal Analysis, Long memory

1. Introduction

It is challenging task to represent many natural phenomena such as rainfall, earthquakes, wind speed, groundwater flow which vary randomly over time with a physical model. Normally natural phenomena exhibit a high degree of randomness, which will not easily show any pattern beside seasonality and trend. Hydrological phenomena are often regarded as principal examples for non-linear systems and apprehended as complex systems. Indeed, most hydrological phenomena are considered as the outcome of simple systems with nonlinear interdependent but sensitive dependence on initial conditions [1]. Studies have shown that the so called random phenomena exhibit some correlations which can be exposed with some analytical tools. Most of these natural phenomena are not subjected to pure chance but exhibit some kind of correlation e.g. [2, 3]. Monotonous trends may lead to an uncorrelated pattern, under the impact of a trend, to look like long-term correlated pattern [4]. Furthermore, it is difficult to distinguish trends from long-term correlations, because stationary long-term correlated time series exhibit persistent behaviour or long memory and a tendency to stay close to the momentary value [5].

Rainfall is one of the challenging components of the hydrological cycle that exhibits a high non-linear and complicated phenomenon and requires standard and well detailed modelling to obtain accurate prediction. The complex nature of rainfall time series has been appreciated for decades, for example, Tiwari & Pandey [6] studied the trend of rainfall long-term record from 1851 to 2006 for seven meteorological regions of India using the methods of Linear trend analysis, innovative trend analysis, sequential Mann–Kendall test and partial cumulative deviation tests. Rakhecha [7] analysed rainwater features using descriptive statistics on seasonal features of rainstorms, areal rainfalls, quantum and rainwater variability that produced droughts and floods in West Rajasthan. He had used rainfall data of 124 years (1871–1994) in a manner that the information became useful in utilising water resources for human activities. Graham & Mishra [8] modelled with 31 years rainfall data (1985–2015) for Allahabad, Uttar Pradesh-India using Box-Jenkins Methodology. Their results indicated that the seasonal Autoregressive Integrated Moving Average model (ARIMA) model provides consistent and satisfactory predictions for rainfall parameters on monthly scale. Uba & Bakari [9] analysed 372 rainfall data observations for the period of 1981–2011 in Maiduguri-Nigeria. Their results indicated that ARIMA (1, 1, 0) provides a good fit for the rainfall data and is appropriate for short term forecast. Olatayo & Taiwo [10] presented a study that utilised emerging Fuzzy Time Series (FTS), ARIMA and the Thiel's regression methods for the analysis and forecasted the dynamical pattern of rainfall occurrences based on historic data.

Rainfalls data modelling is very essential to many hydrological issues, for example, in identification of intense, moderate and low rainfall areas; detecting areas prone to flood, drought, and other hazardous events; and for agricultural purposes [11]. However, most of the literatures deal either with linear or nonlinear modelling approaches e.g. [12, 13]; both approaches achieved successes in their domains. Nevertheless, none of approaches is found to be a common model that is suitable for all circumstances. These problems strengthen our thinking to extract more information from the available rainfall data. In fact, one of the purposes of measuring data is to learn about the mechanisms in the data themselves and to make conclusion about its present and future state.

Thiruvallur is a region of highly variable rainfall in both spatially and temporally. Therefore, the study of rainfall variability is fundamental to examine its impact on socio-economic activities. Rainfall in Thiruvallur is highly seasonal that is nonlinear with an organised pattern of clustered structure and may exhibits multi-scaling features. Understanding the nature of the temporal variability of rainfall is important to improve the predictability of climatic events such as floods and droughts [14]. Thus, it is essential to develop a systematic method that will capture the observed characteristics of the data. This study objectively detects the Thiruvallur rainfall patterns for better understanding of researchers in modelling rainfall time series.

2. Data used and study area

The data used in this study is from the Thiruvallur monthly rainfall records of nine locations such as; Thiruvallur, Korattur_Dam, Ponneri, Poondi, Red Hills, Sholingur, Thamaraipakkam, Thiruvottiyur and Vallur Anicut for the period of over 30 years (1971–2016) collected from Institute of Water Studies, Public Works Department, Government of India. Thiruvallur is one of the fastest developing districts in Tamil Nadu, India. it lies between 12°15′ and 13°15' North latitude and 79°15′ and 80°20′ East longitude. The district experiences semi-arid sub-tropical monsoonal climate. Thiruvallur forms part of Coromondal coastal region, topographically flat with some few hills undulated. The average maximum temperature is between 29°C to 36.6°C with the minimum within 17.3°C to 24.4°C. The average normal rainfall of the district is 1104 mm. Out of this about 50% is received during north east monsoon period and about 40% is received during south west monsoon period (http://www.tnenvis.nic.in. retrieved 02/04/2021). The geographical map of the study area is given in **Figure 1**.

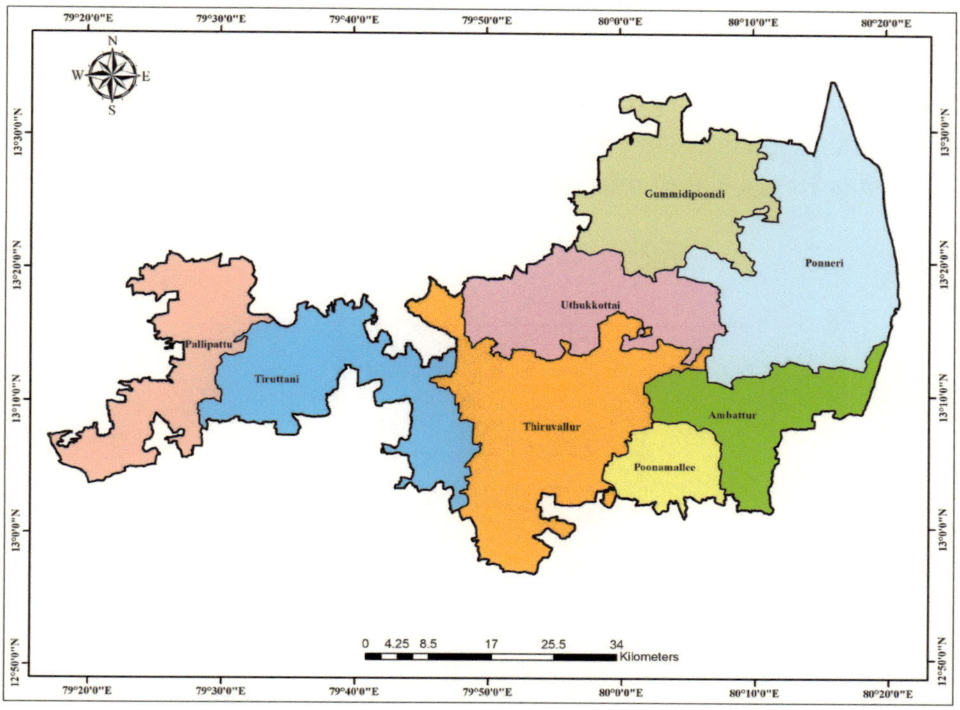

Figure 1.
Geographical map of Thiruvallur district.

3. Methodology

The analysis focus on characterising rainfall based on historical data. The descriptive statistics of the data was first discussed followed by the analysis of fractal scaling properties. The description of the methodology is given in the flowchart in **Figure 2**.

3.1 Fractal scaling analysis

Many geophysical fields appear geometrically complex involving high variability, intermittency and frequent occurrence of extreme values. Fractal scaling analysis, on the other hand presents variety of techniques which can quantify such properties using Hurst phenomenon. The parameter H (Hurst Exponent), display the scaling property of a time series. The Hurst exponent takes values from 0 to 1 ($0 \leq H \leq 1$). If $H = 0.5$, the series is a random walk (Brownian time series) and there is no correlation between any element and a future element, that is; knowing one data point does not provide insight into knowing future data points in the series. If $0.5 < H < 1$, the series indicates persistent behaviour or long memory. In this case, a future data point is likely to be a data point preceding it. If $0 < H < 0.5$, the series is called anti-persistent. In this case, an increase will tend to be followed by a decrease or vice versa [15]. Among the methods used for quantifying the H embraced in this paper are; Detrended Fluctuation Analysis (DFA) and Rescaled Range (R/S) Methods.

3.1.1 Detrended fluctuation analysis (DFA)

Consider a fluctuating time series x_i for $i = 1, 2, 3, \ldots, N$ sampled at equal time interval $i\Delta t$. Assuming that x_i are increments of a random walk process around the mean $\overline{x} = N^{-1}\sum_{i=1}^{N} x_i = 0$, the 'trajectory' of the signal following integration is given as: $y_j = \sum_{i=1}^{j} x_i$, thus, by dividing the profile into distinct segments indexed by $k = 1, 2, \ldots, \left[\frac{N}{n}\right]$, the confined trend is fitted in each segment, by the polynomial $f_k^{(p)}(j)$ of order p, and the profile is de-trended using the expression;

$$Z_j^p = y_j - f_k^{(p)}(j), \text{for}\, j = 1, 2, \ldots, N \tag{1}$$

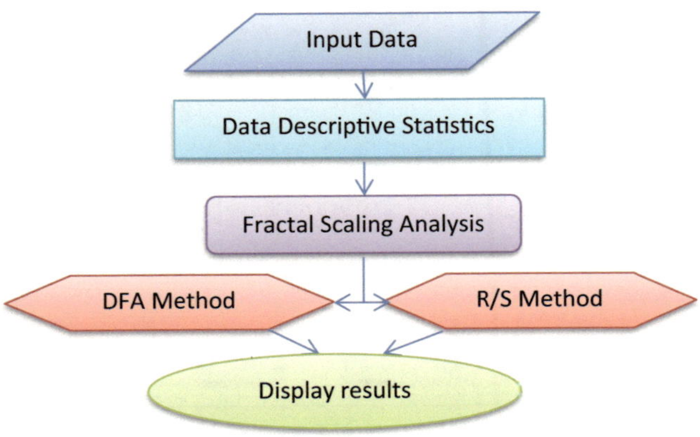

Figure 2.
Flowchart of the methodology.

Following [16], the possible fluctuations can be measured using the root mean square for a given segment of length n.

$$F_p(n) = \sqrt{\frac{1}{n[N/n]} \sum_{j=1}^{n[N/n]} \left(Z_j^p\right)^2} \quad (2)$$

A power-law relationship between $F_p(n)$ and n designates scaling with an exponent δ

$$F_p(n) \sim n^{\delta} \quad (3)$$

and such a process has a power-law autocorrelation function $C(\tau) = \langle x_j x_{j+r} \rangle \sim \tau^{-\alpha}$, where $0 < \alpha < 1$, and the relationship between the correlation exponents is $\alpha = 2(1 - \delta)$ [17]. Hurst exponent H can be obtained directly from the scaling exponent α, that is, $\alpha = H$.

3.1.2 Rescale range method

The Rescale Range formula is given as [18]:

$$(R/S)_n = cn^H \quad (4)$$

where $(R/S)_n$ represent the rescaled range (R/S) statistic measured over a time index n, the terms c and H represents a constant and the Hurst exponent respectively. The estimation of the Hurst exponent is done by taking the logarithm of (4) to give:

$$log\,(R/S)_n = \log(c) + Hlog\,(n) \quad (5)$$

H can be estimated as the slope of log/log plot of $(R/S)_n$ versus n.

Consider a rainfall time series x_t, for $t = 1, 2, \ldots N$, the (R/S) can be defined as the series representing cumulative deviations from the mean of the rainfall series rescaled by its standard deviation. In the *(R/S)* logic, long memory or long term dependence is considered as the extended periods of whole similar behaviour with unequal duration. The methodical process to estimate $(R/S)_n$ values can be described in the steps below following [15]:

Step1: The time period of a time series of length N is grouped into m adjoining sub groups of length n such that $m \times n = N$ with sub group carrying x_{ij} where $i = 1, 2, \ldots, n$ denotes the number of terms in each sub group and $j = 1, 2, \ldots, m$ denotes the sub group index. For each sub group j the (R/S) statistic can be estimated as:

$$(R/S)_j = S_j^{-1}\left[max_{1 \le k \le n} \sum_{j=1}^{k} (x_{ij} - \overline{x}) - min_{1 \le k \le n} \sum_{j=1}^{k} (x_{ij} - \overline{x})\right] \quad (6)$$

where S_j denote the standard deviation for each sub group. In Eq. (5), the kth deviations from the sub group mean have mean equal to zero, therefore the last value of the cumulative deviations for each sub group will equally be zero. Hence, the maximum value of the cumulative deviations will be greater than or equal to 0, whereas the minimum value of the cumulative deviations will be less than or equal to 0. Thus the bracketed term, that is, the range value will be non-negative.

Step2: The $(R/S)_n$ is computed by taking the average of $(R/S)_j$ for all the m adjoining sub groups with length n resulting to:

$$(R/S)_n = \frac{1}{m}\sum_{j-1}^{k}(R/S)_j \tag{7}$$

Step 3: Note that Eq. (6) calculates the *(R/S)* value corresponding to a certain groups of length n. While applying Eq. (5), steps 1 and 2 are repeated by increasing n values until $n = N/2$. After these steps, it is apparent that the time width is contained in the *(R/S)* analysis by examining whether the range of the cumulative deviations depends on the length of the whole time period. After Eq. (7) is estimated for different n periods, the Hurst exponent can be estimated through an ordinary least square regression from Eq. (5).

4. Results

4.1 Descriptive statistics of the rainfall data sets

The descriptive statistic of the considered 9 monthly rainfall series is given in **Table 1**. It could be observed from the table that S01-S08 series follow the same statistical patterns with standard deviation less than the mean and S09 has standard deviation greater than the mean. The Coefficient of Variation (C.V) measure the distribution of data points around the mean. It symbolises the ratio of the standard deviation to the mean. Data with a C.V value less than 1 is considered to have low-variability, while that with a C.V value higher than 1 is considered to have high variability [19]. From **Table 1**, the C.V's for all the data sets are higher than 1 which indicate that the rainfall fluctuates significantly through time except that of S09 which shows negligible C.V from the mean.

4.2 Fractal scaling analysis

Figure 3a–i depicted the results of the (DFA) with fractal scaling properties. DFA gives estimates of the degree of detrended fluctuations at different periods (window size t) of the rainfall time series. It measures the scaling association between estimates and the window size. The estimation of the Hurst parameter H by (DFA) method shoulders self-similarity in the rainfall series. The signal is said to be self-similar if the detrended fluctuations increases as a power law function of time scale and yield a straight line on a log–log fluctuation plot as the association

Station	Station code	Mean	Std. dev.	Skewness	Kurtosis	C.V
Thiruvallur	S01	108.74	139.99	2.05	8.33	1.29
Korattur_Dam	S02	102.24	138.23	2.49	12.79	1.35
Ponneri	S03	114.63	160.33	2.23	9.52	1.39
Poondi	S04	101.11	133.30	2.89	18.93	1.32
Red Hills	S05	117.61	174.87	2.88	15.21	1.49
Sholingur	S06	73.30	94.92	2.08	9.14	1.28
Thamaraipakkam	S07	99.55	128.83	2.13	9.97	1.29
Thiruvottiyur	S08	77.85	115.49	2.54	11.38	1.48
Vallur Anicut	S09	113.64	110.19	0.47	1.77	0.97

Table 1.
Summary of the descriptive statistics of the rainfall data sets.

between the estimates and window size t. The slope of the plot is the scaling exponent estimate which gives the fractal scaling property also summarised in **Table 2** for the 9 locations.

The statistical technique based on (R/S) is designed to assess the nature and extent of variability in data over a time period with the purpose of providing an assessment of how the apparent variability of the rainfall series changes with the length of the time-period 1971 to 2016. (R/S) reveals whether or not a time series exhibits persistence or anti-persistence. A log–log plot of the (R/S) statistic versus the number of points of the aggregated series (**Figure 4a–i**) formed a straight line

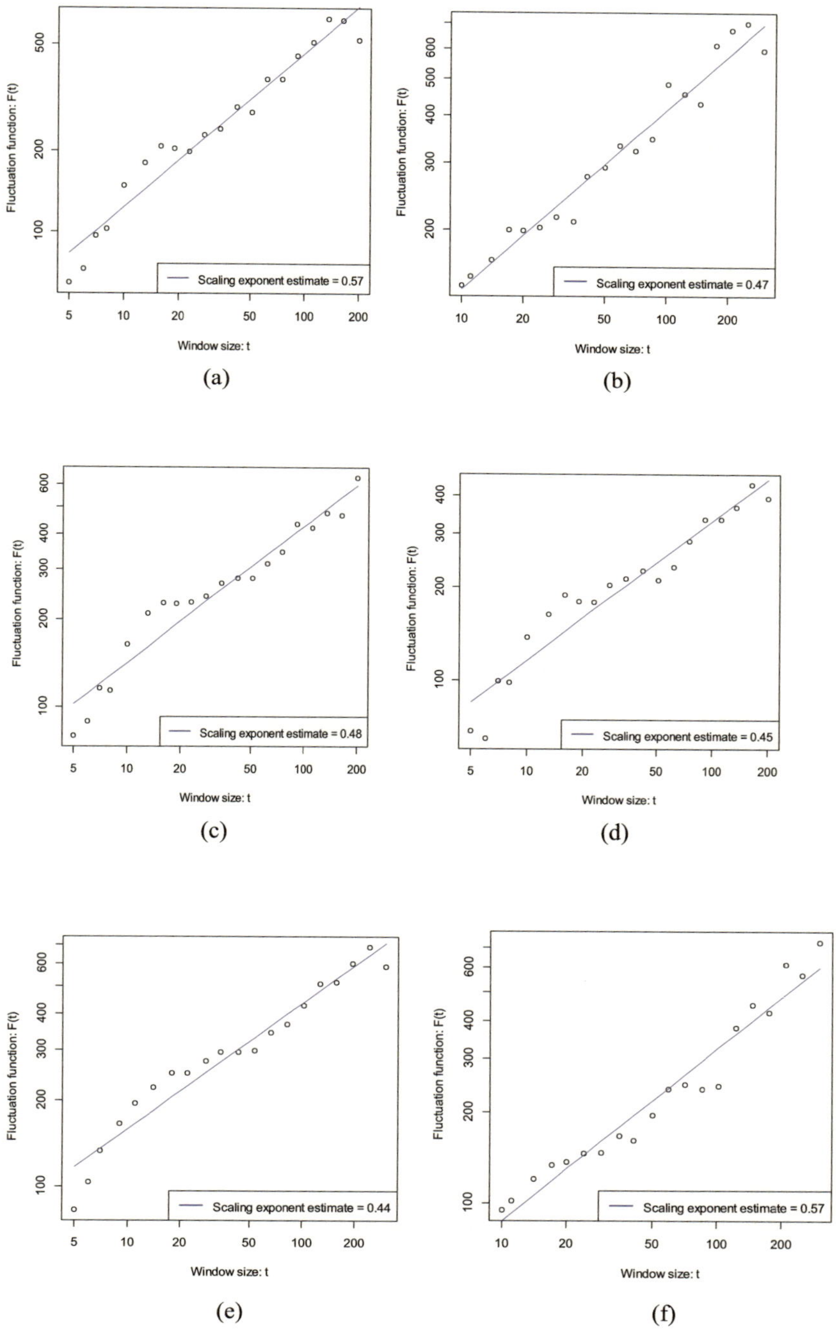

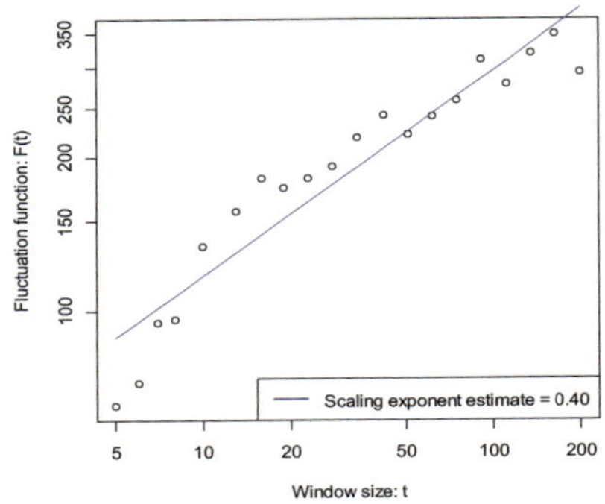

(g)

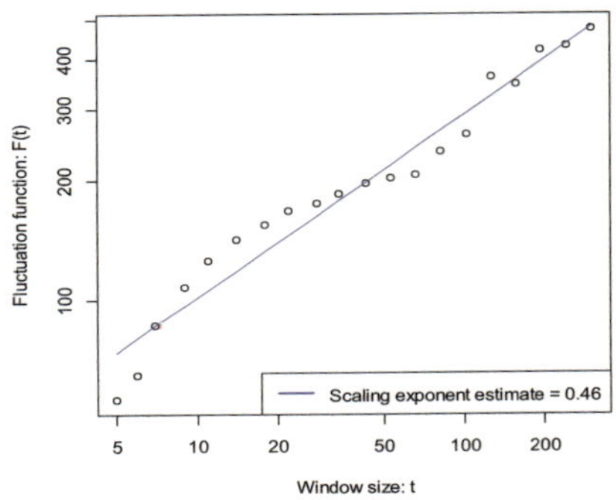

(h)

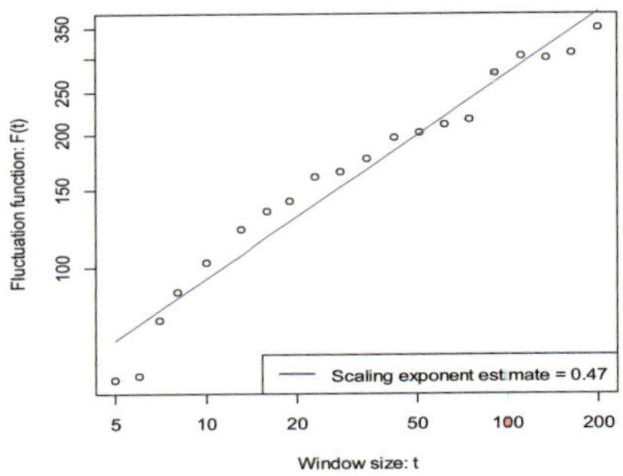

(i)

Figure 3.
Hurst exponents results based on DFA method for (a) Thiruvallur, (b) Korattur Dam, (c) Ponneri, (d) Poondi (e) Red Hills, (f) Sholingur, (g) Thamaraipakkam, (h) Thiruvottiyur, and (i) Vallur Anicut.

Station	Station code	DFA	(R/S)
Thiruvallur	S01	0.57	0.58
Korattur_Dam	S02	0.47	0.44
Ponneri	S03	0.48	0.52
Poondi	S04	0.45	0.51
Red Hills	S05	0.44	0.42
Sholingur	S06	0.57	0.56
Thamaraipakkam	S07	0.40	0.40
Thiruvottiyur	S08	0.46	0.45
Vallur Anicut	S09	0.47	0.48

Table 2.
Summary results for Hurst exponents based on detrended fluctuation analysis and rescaled range methods.

with the slope being an estimate of the Hurst parameter value. The (R/S) results are summarised in **Table 2**.

The evidences in **Figures 3** and **4** for *(R/S)* and DFA methods clearly shows that monthly rainfall time series of Thiruvallur and Sholingur have Hurst exponent values within $0.5 < H < 1$ which is an indication of strong persistence or long memory, as such the series have a predictable component. Thamaraipakkam, Thiruvottiyur, Vallur Anicut, Korattur Dam and Red Hills have H values less than 0.5, indicating a property called anti-persistent where in an increase will tend to be followed by a decrease or vice versa [14]. Ponneri and Poondi have conflicting results based on the two methods, however, their H values are approximately 0.5 showing a random walk behaviour, and in this, there is no correlation between any part of the data. That is, knowledge of one data point does not provide insight to predict future data points in the series.

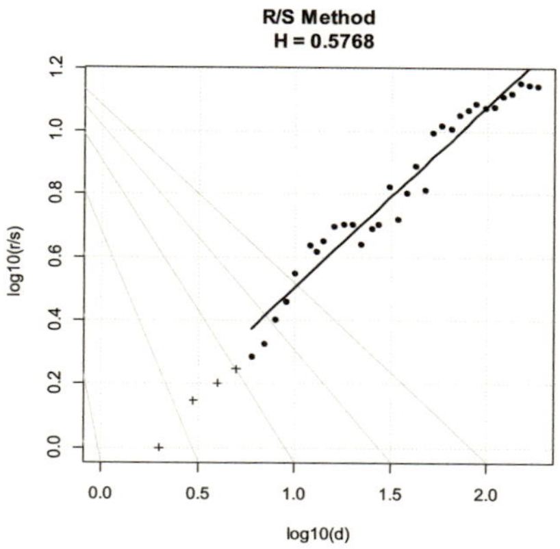

(a)

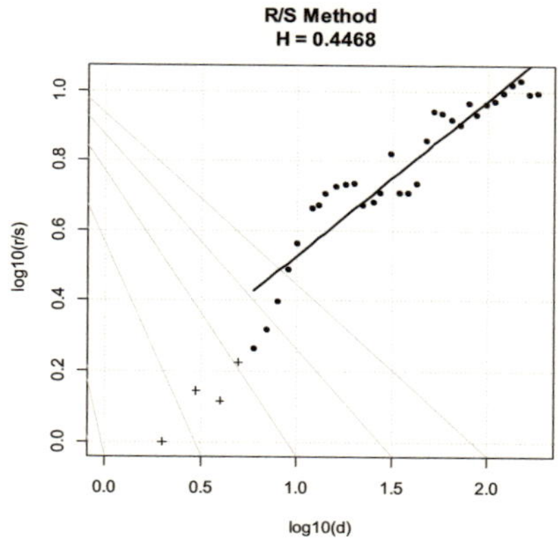

(b)

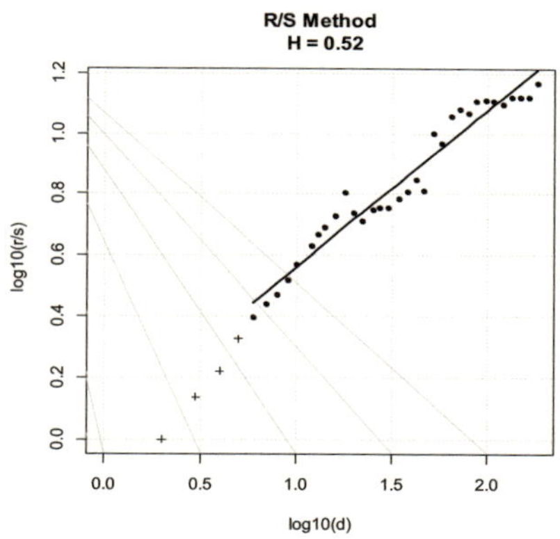

(c)

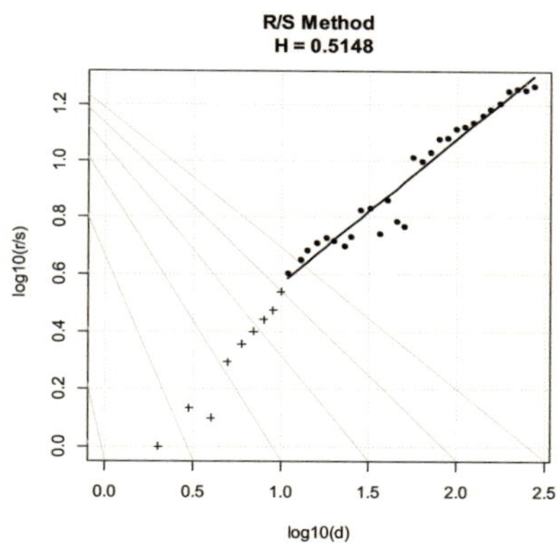

(d)

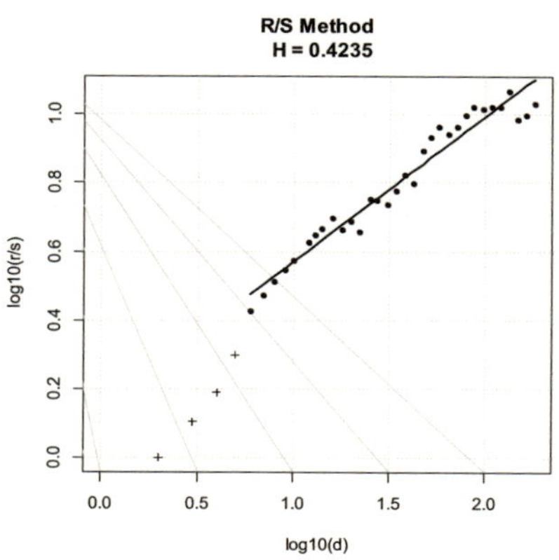

(e)

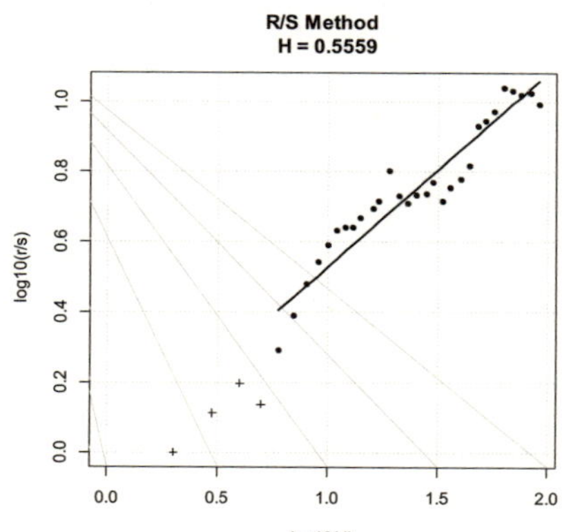

(f)

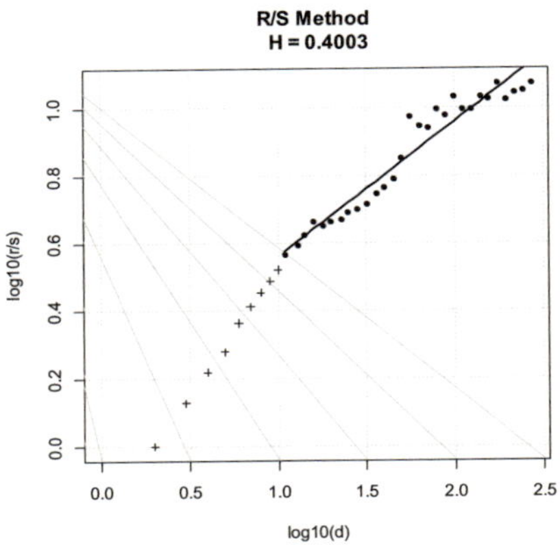

(g)

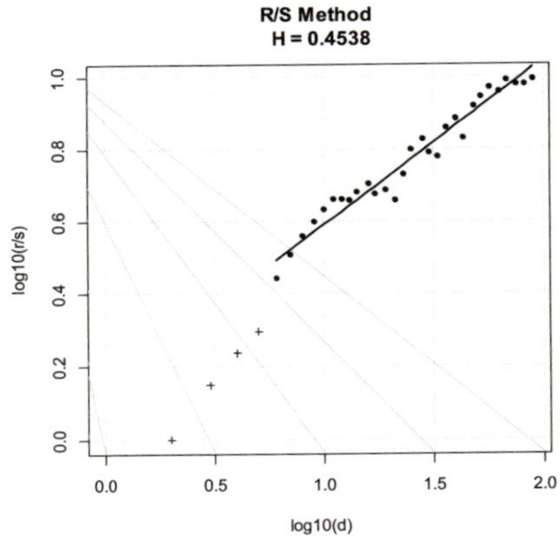

(h)

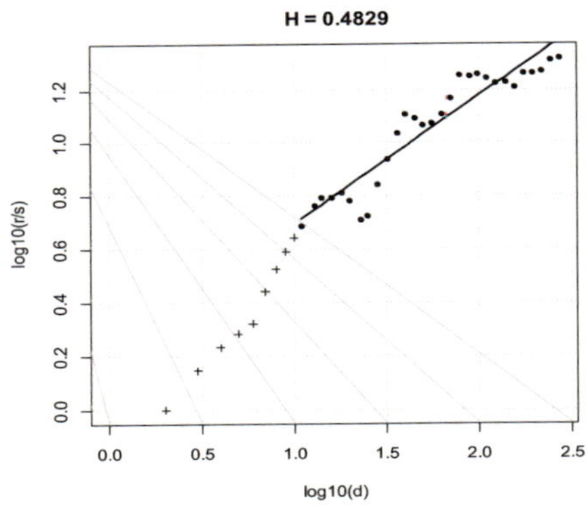

(i)

Figure 4.
Plots of (R/S) Hurst estimates for (a) Thiruvallur, (b) Korattur Dam, (c) Ponneri, (d) Poondi (e) Red Hills, (f) Sholingur, (g) Thamaraipakkam, (h) Thiruvottiyur, and (i) Vallur Anicut.

5. Conclusion

Monthly rainfalls in different locations of Thiruvallur district exhibit a tendency of randomness in the long run. The presence of a changing deterministic pattern was examined through a method that allows detecting both apparent and hidden features in rainfall time series. The fractal Scaling analysis based on DFA and the (R/S) methods reveals typical behaviours shown by all the rainfall time series, some are persistent and purely random, some behaves as random walk and some have anti-persistent behaviour. This shows that there is no universal model for predicting rainfall in Thiruvallur district. Rather, rainfall in a location need to be treated based on its associated features. Non consideration of fractal features in hydrological variable modelling may lead to spurious estimates. Finding appropriate model to estimate and predict future rainfalls with consideration to the observed characteristic would be a subject for future research.

Conflict of interest

The authors declare no conflict of interest.

Author details

Ibrahim Lawal Kane[1*] and Venkatesan Madha Suresh[2]

1 Department of Mathematics and Statistics, Umaru Musa Yar'adua University, Katsina, Nigeria

2 Director, G.S.Gill Research Institute and Professor and Head (former) Centre for Natural Hazards and Disaster Studies, University of Madras, Chennai, India

*Address all correspondence to: ibrahim.lawal@umyu.edu.ng

References

[1] Sivakumar, B. (2004). Chaos theory in geophysics: past, present and future. Chaos, Solitons and Fractals. 19, 441-462.

[2] Gil-Alana, Luis A. (2012). U.K. Rainfall Data: A Long-Term Persistence Approach. J. Appl. Meteor. Climatol. 51, 1904–1913.

[3] Lovallo, M., Shaban, A., Darwich, T. and Telesca, L. (2013). Investigating the time dynamics of monthly rainfall time series observed in northern Lebanon by means of the detrended fluctuation analysis and the Fisher-Shannon method. Acta Geophysica, DOI 10.2478/s11600-012-0094-9.

[4] Nnaji C. C. (2011). Time Series Analysis of Monthly Rainfall In Nigeria With Emphasis on Self-Organized Criticality. Journal of Science and Technology.Vol. 31, No. 1, pp 139

[5] Koscielny-Bunde, E. Kantelhardt, J. W. Braun, P. Bunde, A. and S. Havlin, S. (2006). Long-term persistence and multifractality of river runoff records: Detrended fluctuation studies. J. Hydrol. 322, 120– 137.

[6] Tiwari, H., and Pandey, B.K. (2019) Non-parametric characterization of long-term rainfall time series. Meteorol Atmos Phys 131, 627–637. https://doi.org/10.1007/s00703-018-0592-7

[7] Rakhecha PR. (2018). Hydrometeorology of the Rajasthan desert rainfall. Int J Hydro.; 2(6):688–694. DOI: 10.15406/ijh.2018.2.00144

[8] Graham A, & Mishra E. P. (2017). Time series analysis model to forecast rainfall for Allahabad region. Journal of Pharmacognosy and Phytochemistry; 6(5): 1418-1421. E-ISSN: 2278-4136.

[9] Uba E. S & Bakari H. R. An Application of Time Series Analysis in Modeling Monthly Rainfall Data for Maiduguri, North Eastern Nigeria. Mathematical Theory and Modeling, 2015, Vol.5, No.11, Pp 24-33. ISSN 2225-0522 (Online), www.iiste.org.

[10] Olatayo T. O, & Taiwo A. I. (2004). Statistical Modelling and Prediction of Rainfall Time Series. Global Journal of Computer Science and Technology, 2014, G Interdisciplinary Volume 14 Issue 1 Version1.0. (USA) Online ISSN: 0975-4172 & Print ISSN: 0975-4350

[11] Patakamuri S.K, Krishnaveni M. and Venkataramana S. (2020). Long-Term Homogeneity, Trend, and Change-Point Analysis of Rainfall in the Arid District of Ananthapuramu, Andhra Pradesh State, India. Water, 12, 211; doi:10.3390/w12010211.

[12] Kokilavani S, Pangayarselvi R, Ramanathan SP, Dheebakaran Ga, Sathyamoorthy NK, Maragatham N & Gowtham R.(2020). SARIMA Modelling and Forecasting of Monthly Rainfall Patterns for Coimbatore, Tamil Nadu, India. Current Journal of Applied Science and Technology, 2020; 39(8): 69-76, Article no.CJAST.55802.

[13] Csaba I, Endre T. and Péter S. (2018). Examination of rainfall data for 110 years using spectral and wavelet analysis, Central European Geology, Vol. 61, 1, 1–15. DOI: 10.1556/24.61.2018.01

[14] Morata A, Martı'n M L, Luna M Y, & Valero F. Self-similarity patterns of precipitation in the Iberian Peninsula. Theoretical and Applied Climatology. 2006, 85, 41–59. DOI 10.1007/s00704-005-0175-7

[15] Alptekin, N. (2006). Long memory analysis of USD/trial exchange rate. Int. J. Soc. Sci., 1, 111–112.

[16] Peng, C.K., Buldyrev, S. V. and Havlin, S. (1994). Mosaic organization

of DNA nucleotides. Phys. Rev. E. 49 (2), 1685 – 1689.

[17] Koscielny-Bunde, E., A., Bunde, S., Havlin, H.E., Roman, Y., Goldreich, H. and Schellnhuber, J. (1998). Indication of a universal persistence law governing atmospheric variability. Physics Review Letters. 81, 729-732.

[18] Talkner, P. and Weber, R.O. (2000). Power spectrum and Detrended fluctuation analysis: application to daily temperatures. Phys Rev. E62, 150–160.

[19] Chandramouli, A. (2013). The future of U.S manufacturing: a change manifesto. Retrieved on November 23, 2013, from www.cognizant.com, Last access: 23/11/2013.